DREAMERS
and DOERS
Sailing the South Pacific

ARLENE GALISKY

DREAMERS AND DOERS
SAILING THE SOUTH PACIFIC

iUniverse books may be ordered through booksellers or by contacting:

iUniverse
1663 Liberty Drive
Bloomington, IN 47403
www.iuniverse.com
1-800-Authors (1-800-288-4677)

Because of the dynamic nature of the Internet, any web addresses or links contained in this book may have changed since publication and may no longer be valid. The views expressed in this work are solely those of the author and do not necessarily reflect the views of the publisher, and the publisher hereby disclaims any responsibility for them.

Any people depicted in stock imagery provided by Getty Images are models, and such images are being used for illustrative purposes only. Certain stock imagery © Getty Images.

ISBN: 978-1-5320-5185-2 (sc)
ISBN: 978-1-5320-5186-9 (e)

Library of Congress Control Number: 2018907700

Print information available on the last page.

iUniverse rev. date: 07/09/2018

For David,
Together, we turned a dream into reality.

Contents

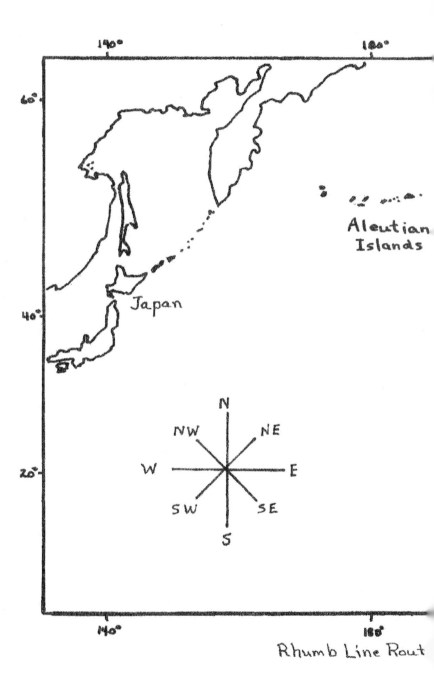

Rhumb Line Rout

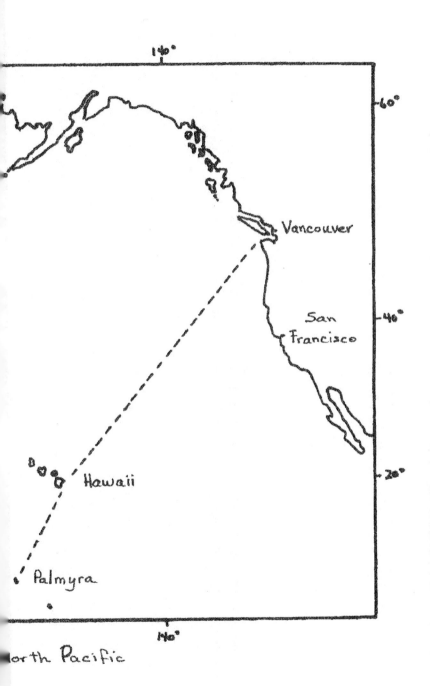

North Pacific

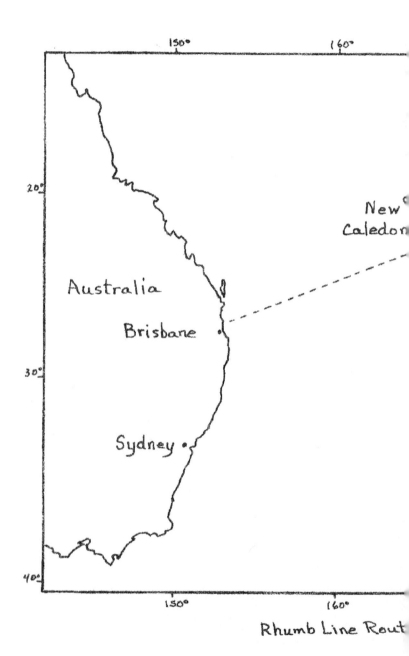

Rhumb Line Rout

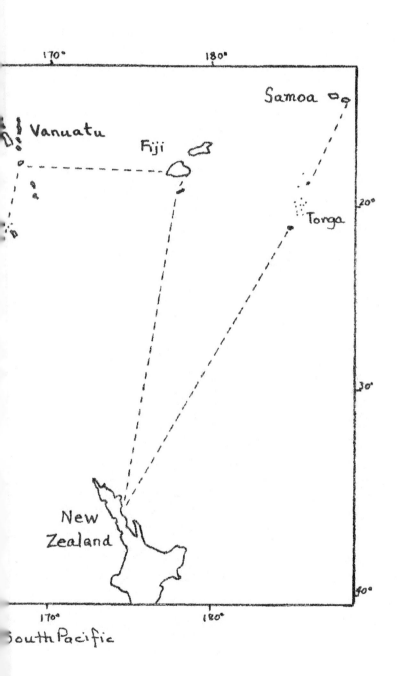

South Pacific

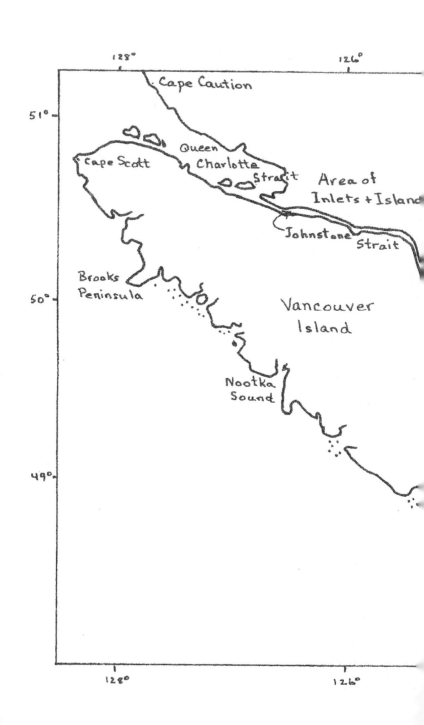

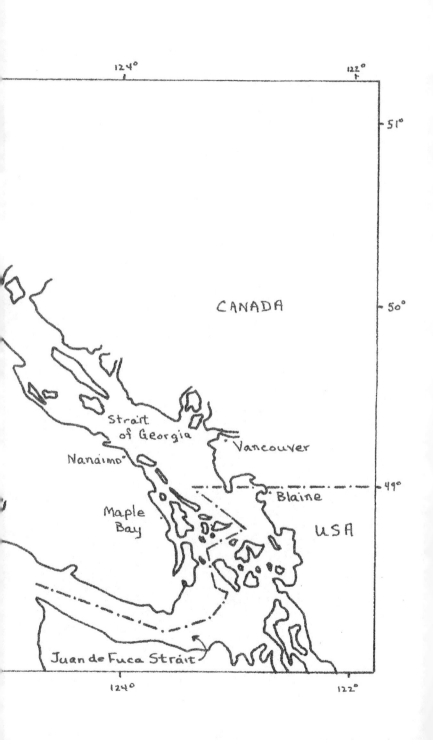

After the haulout in New Zealand

Saturday, June 8, 1996
Victoria, B.C.

PASSAGE PLAN

FOR SAILING VESSEL, WINDY LADY III
Description:
  40-foot cutter-rigged sloop monohull
  flush deck, aft cockpit with pilot house
  white with blue trim.
Port of Registry: Vancouver, Canada
Registration #804823.
Registered tonnage: 13.97

TRAVELLING TO HILO, HAWAII
  Departing Victoria Harbour on June 8, 1996 proceeding
through Strait of Juan de Fuca to Cape Flattery. Will stop
at Neah Bay only if weather dictates.
  Expecting to continue directly to Hilo, Hawaii.
  Intended route is 100 miles off the west coast down to San
Francisco, then west to Hawaii. Depending on winds, the
track will be approximately:
  SW to 130 degrees W, 45 degrees N
  S to 130   "   W, 35   "   N
  SW to 150   "   W, 25   "   N
  SW to 155   "   W, 20   "   N (Hilo)

FUEL ON BOARD: 750 liters of diesel
RANGE: 1200 NM

WATER: 900 liters
FOOD: Provisions for 24 weeks

RADIOS AND ELECTRONIC EQUIPMENT: Call sign CFL 4350
  VHF, ICOM IC-M58
  SSB, ICOM IC-700
       HAM VE0-DLB
  Radar, Furuno model 1721 (24 mile)
  GPS, Furuno GP-50 Mark 2

SAFETY EQUIPMENT
  RFD 4-man life raft
  8.5 foot white rigid dinghy on foredeck
  EPIRB, Lo-kata 406 (A78D00800500401)

S.O.B. David L. Ball, Captain
        Arlene L. Galisky, Navigator
        Brian O. Ball, Crew

EXPECTED TIME ON ROUTE: 24 days

FLOAT PLAN FILED WITH: Gordon Ball

CHAPTER 1

The Dream

When darkness fell, we were still hours away from the Australian coast; the wind had died again and we were motoring. We spotted lights on shore about midnight, appearing and disappearing as *Windy Lady* rose and fell in three-foot seas. An hour later, we were anxiously searching for the red and green lights marking the entrance to the shipping channel into Moreton Bay. When the sun later brightened the sky over Moreton Island, we would see the Queensland coast for the first time.

Dave and I were within hours of achieving our dream of sailing from Victoria, Canada, to Brisbane, Australia. The dream was only three years old, but the time had been tumultuous. We'd bought a sailboat and learned to sail, then prepared the boat and ourselves for a voyage across the Pacific Ocean. Sailing out of Victoria harbor seventeen months earlier, we'd spent 110 days at sea, and visited Hawaii, Palmyra, American Samoa, Tonga, New

Zealand, Fiji, Vanuatu and New Caledonia. Now, before us, was Australia!

It was an unlikely accomplishment for two people whose only familiarity with the ocean came from a couple of winter holidays in Hawaii. Dave came up with the plan; he was the dreamer. Four years earlier, he'd been a partner in a public accounting firm in Prince George, a small city in central BC. He had grown restless, however, and lost interest in the work. With their two children grown, he and his wife were also in the process of separating.

His retirement funds, which had limped along for decades, had started to grow a few years earlier. One day he realized that with some careful planning, he could retire at age fifty-five. While he welcomed the idea of retirement, he also worried. He'd had partners who retired and were dead within a year. After going into the office every day for years on end, what could he do differently?

Dave was a big man, standing over six feet tall and weighing two hundred pounds; he was still vital and healthy. He had grown up in southern Saskatchewan, and by the time he graduated from high school, had worked as a farm hand, a section hand on the railroad, and a roughneck on an oilrig. In the years that followed, he joined the air force, worked in an oil refinery, and immigrated to Australia, where he worked underground in a mine for six months. He then returned to Canada and articled with an accounting firm.

He just wasn't the type to sit back and take it easy, so now thought about buying a 4x4 pickup and driving from BC to the southern tip of South America. He lost interest, however, after reading about the Darien Gap and other difficulties faced by travelers in Central America. He didn't have to look far for another idea, as he was then receiving the occasional letter from a friend, who was sailing around the world. Although he'd never been on a sailboat, it wasn't a difficult leap from a truck to a boat and from South America to the South Pacific.

He began studying cruising magazines and researching sailboats; soon he knew the *Boat Trader* inside out. His resolve was shaken by his father's death in November 1993, but his mother's passing the following July spurred him into action. With only a month remaining until he retired, he made up his mind and came to see me. "Lene," he declared, "I'm going to buy a sailboat and sail to Australia." Then he added quietly, "Do you want to come?"

Dave and I had been friends for years, often hiking or whitewater canoeing together, sharing an enthusiasm for outdoor adventure. At one time, admittedly, such an invitation would have been the answer to my prayers, but that time had long since passed. I was forty-eight years old with my own career, family obligations, and community involvements. As it happened, I was also at a turning point in my life.

Five years earlier, I'd become active in a local campaign to save a river from a diversion project. I felt passionately about it and became so involved that it turned into a second full-time job. An environmental review panel had finished hearings just the previous week, and now that it was over, I was exhausted and despondent. I needed time off, probably had to make some changes, but wasn't up to thinking about it yet.

As I listened to Dave reveal his plans, I knew that he was not talking about a simple change. To do this would mean walking away from everything I had achieved, as well as any ambitions I had for the future. Still, I said, "Let me think about it for a day or two."

I had just started to do so when a memory popped into my head of an event that occurred twenty years earlier. I'd been living in the west coast city of Vancouver and was walking with a friend in Stanley Park. We had stopped to watch the sun setting over the Strait of Georgia, and immersed in the view, I was startled to hear my friend ask, "What was that?" Only then realizing that I'd spoken my thoughts aloud, I nodded my head towards the sea and repeated firmly, "One day, I'm going out there!"

Maybe those words simply reflected a hidden yearning for adventure or freedom; maybe they didn't mean anything at all, as I hadn't thought of them once in the years since. But I knew instantly that I could never pass up such an adventure. There was no decision required, and within days, I'd given three months' notice to my employer.

As neither Dave nor I had sailed before, we first signed up for a course offered by a sailing school in Vancouver. In early September, we joined an instructor and one other couple onboard a thirty-four-foot sailboat and spent five days cruising in the Strait of Georgia, a 240-km long arm of the ocean lying between Vancouver Island and the mainland of BC. As winds were light, we mostly just raised and lowered sails, but did motor through fog and ran aground on a sunny afternoon. As neither of us showed any signs of being seasick, our plans moved forward.

By the end of the month, Dave had found his boat. *Windy Lady* was a beamy, forty-foot, fiberglass sloop with a rear cockpit shielded only by a windscreen. She had large windows in the main cabin, high ceilings, and more living space than other boats we'd inspected. Significantly, banging our heads against bulkheads wasn't nearly the problem that it was on some boats.

We took possession in early October and arranged to spend the preceding night onboard, as we wanted a full day to move the boat from Blaine, Washington, to Maple Bay on Vancouver Island. In the morning, we awoke to thick fog and then had to wait for the venders, Ron and Diane, to pick up the last of their belongings. By the time we untied the mooring lines, it was almost noon.

As we motored out of the harbor, we saw Ron standing on the edge of a pier, waving his arms. Above the throbbing of the diesel motor, I barely heard him yell, "*Engine Water!*" That meant nothing to me but sent Dave scurrying below. Raising the floor (sole) over the engine compartment, he opened a valve that allowed seawater to flow into the cooling system. Our education had begun.

Steering the boat westward, across the Strait of Georgia, Dave headed towards a string of small islands some forty km away. The afternoon was perfect, with a warm, sunny sky and rippled sea, and we reveled in every minute of the five-hour crossing. I did feel alarmed for a few minutes when I realized that the pass through the islands wasn't shown on our chart, but the frequent appearance of BC Ferries at the entrance soon calmed my fears.

By the time we were through the pass, the sun was low in the sky. As we were only halfway to our destination, we opted to stop at a harbor on nearby Saltspring Island. The sky was growing dark when we tied up at the main dock, and Dave immediately located a phone and called Canada Customs. He had asked Ron about crossing the US-Canada border and been assured that it wasn't a problem. Ron had explained, "I do it all the time. You just have to phone in as soon as you arrive."

When Dave checked in, however, that was not the message he heard. An angry agent barked into the phone, "You bring that boat to the Customs Dock in Victoria *now*!" As we simply weren't equipped to travel after dark, he finally relented but issued a stern warning, "Everyone's entitled to one mistake, and you've had yours!"

We found our way to Maple Bay Marina the next day, and our plans quickly moved forward. By the end of October, I had worked out my notice and we were ready to move onboard. Probably because of our own uncertainties, we were slow to tell friends and family what we were up to; when we did, reactions varied. At one end of the spectrum was my mother, who asked, "Why on earth would you leave a perfectly good house to go live on a boat?" At the other end was Dave's younger brother, Brian, who stood solidly with us. He was also a dreamer, having crewed on two ocean passages, and became a sounding board for all the issues that we needed to resolve.

The winter of 1994-95 proved to be cold and damp onboard *Windy Lady*. In truth, the boat was not fit to live in, let alone take to sea, and I have to wonder how we survived with our plans intact. The cabin windows leaked when it rained, as did the Plexiglas instrument panel in the cockpit, which funneled water onto the foot of the starboard berth. The main water tank had a leak, as did both the hot water tank and the sink in the head. Water then accumulated in the bilge, along with oil from an engine leak.

There were three batteries onboard, all of which needed replacing. One was dedicated to the engine, and the other two were large 8D's with a combined capacity of 450 amps. They made up the house bank and provided an onboard source of power for lights and instruments. The boat was wired for shore power, too, and I had my first lesson in energy management within hours of moving aboard. I plugged in the coffee maker and a small ceramic heater at the same time, blowing the circuit breaker on the dock, which were limited to 20 amps.

We had two fuel tanks, two water tanks and a holding tank in the head. When the holding tank overflowed, the whale pump that was supposed to pump it out didn't work. Then, when the nights grew colder, we woke up with water dripping on our heads from condensation that formed on the exposed fiberglass. The furnace, which should have dried the boat out as well as warmed it up, turned out to be an orphan and never did work. Fortunately, Brian visited regularly and his encouragement and enthusiasm helped keep the dream alive when reality became a little too discouraging.

Dave hired a diesel mechanic in early November, and while we waited for him to finish another job, we set to work. We first cleaned up and equipped the galley, then went through the boat, emptying and cleaning lockers. We also rolled out the inflatable dinghy and found a nail-sized hole near the bow. We began making lists: one for all the jobs we had to do; another, for boat supplies, parts and tools; a third, for food and personal items, including dental appointments and vaccinations. In fact, before we completed one list, another had taken its place, and we threw the last one overboard when we finally went to sea.

After a quick look through the boat, Ian, our diesel mechanic, concentrated on the electrical panel, where a mishmash of different colored wires led in all directions. He and Dave traced all the wires, labeling them, so we knew which items connected to which circuit breaker. He then moved on to the array of hoses that ran between engine block and heat exchanger, as well as fore and aft on the boat. Some were part of the engine's salt-water cooling system; others belonged to the fresh-water system that heated the hot-water tank and cabin radiators. I made sketches of everything.

As Ian worked his way through the diesel engine and transmission, he explained everything he could think of about anything that might be useful. He told us we needed a secondary fuel filter, and that regulations required a propane alarm be installed. We also learned about the driveshaft packing-gland and the three zinc anodes, which protected expensive metal parts from galvanic corrosion. One was in the heat exchanger and two were on the propeller shaft, and they had to be replaced annually. When the boat was ready for sea-trials, he accompanied us out into the bay, but we couldn't raise the mainsail as all four winches were seized up.

Because we were in constant need of various bits and pieces, we opened a charge account at a chandlery across from the dock. The owner, Jim, proved a big help when it came to dealing with unfamiliar products and equipment. We then ordered a new mainsail cover from a local sail maker, and had him cover new foam mattresses for the stern berths. Eventually, we had him make a canopy that shaded most of the deck surface.

As the weeks passed, we adjusted to living onboard, learning to recognize normal boat movements caused by winds and the ten-foot tides, as well as the sounds of water pump and refrigerator. We learned practical lessons, such as the cost of moorage, which depended on the length of the boat and was considerably more for a fifty-footer than a forty-footer. In due course, Dave hired a local carpenter, who finished the two stern berths, put shelves over the galley and chart tables, and installed grab rails throughout the cabin. By then, we'd learned that *BOAT* actually stood for *"Bring Out Another Thousand."*

We didn't do much sailing that winter as quirky winds in the islands frequently died, leaving us to return home under power. So, the reality of sailing a fourteen-ton, forty-foot vessel really hadn't started to sink in. We did learn, however, that maneuvering *Windy Lady* at low speeds in the marina was no easy task. With six feet of hull below the waterline for much of her length and a fair amount of windage at the bow, even light winds and currents affected her. There was also prop-walk, which kicked the stern sideways to port when forward gear was engaged; in reverse, it did the opposite. Of course, the bow went in the opposite direction.

By the beginning of May, we had completed many of the jobs on our "to do" lists, and had installed GPS, radar, and VHF and HF radios. We had also purchased new batteries, sails, lines, and anchor chain. As we wanted to leave Canada by the end of July but still had much to learn about handling the boat, Dave now decided to spend six weeks cruising up the BC coast.

CHAPTER 2

A Reality Check

After thinking about how to make the best use of the time, Dave concluded that we should circumnavigate Vancouver Island. By traveling north up the Inside Passage and then sailing offshore on the return journey, he figured we would gain all the experience we needed. We would initially spend a few days sailing in the Strait of Georgia, checking out procedures and equipment, then stop in Nanaimo, where we could sort out any last minute snags before heading north.

Vancouver Island is 460 km long and 100 km across at its widest point, so suddenly, we were planning to sail over 1,300 km. However, measurements at sea are neither metric nor imperial. Distance is measured in nautical miles (nm), and boat speed and wind speed are measured in knots (kt). Both are roughly twice the value of their metric equivalents, so our voyage would take us about 700 nm.

Setting a departure date of May 13, we worked feverishly, completing a few more jobs and buying and stowing all the supplies we could think of. We were still picking up items at the chandlery the day we left, so made a Hudson Bay start, untying about midday and not going far. Although we started out under sail, the wind died after half an hour, so we motored most of the way to Princess Cove on Wallace Island.

With Dave at the helm, I relaxed for the first time in weeks, but then grew anxious as I thought of the steep learning curve ahead. I was particularly concerned about anchoring for the first time. Although I knew the procedures to follow, we'd never really discussed how we would go about it. Dave obviously had no such concerns, however, and confidently steered *Windy Lady* into the cove, passing by five boats already at anchor. Their crews were sitting out on deck, enjoying the late afternoon sunshine, and I noticed that each boat had a stern line leading to a rocky outcropping on shore.

As I looked around, I grew more uneasy; the cove was long and narrow, and the five boats seemed to take up a lot of space. But Dave just picked a spot, brought the boat to a stop with the stern facing the rock wall, and shoved the gearshift lever into reverse. He then called, "Lene, you want to come and take the helm?"

That caught me completely off guard because Dave always drove the boat in these situations. The problem was that I couldn't push the gearshift lever into reverse, so really couldn't maneuver it. The thought of trying to do so now sent a wave of panic through me, and suddenly, the cove seemed even smaller, while the boat had grown to the size of the *Queen Mary*.

If I wasn't going to drive the boat, then I had to drop the anchor. So, I walked up to the bow and stood looking down at the windlass, which was the winch used to raise the anchor. It was a foreign looking piece of equipment bolted to the deck, and I didn't have a clue how it worked. Worse yet, I didn't know where to start. The fact was that I was never interested in things mechanical. When we were young, my brother had taken clocks and radios apart, but I hadn't been the slightest bit curious. Looking desperately at the dark waters around the boat, I now choked back another wave of panic.

It turned out the windlass was simple to operate, and after a few terse words from Dave, I managed to figure it out. In the process, I dropped the anchor off the bow three or four times, but grew a little more desperate each time, as the chain clang-clanged loudly on the way out, then clang-clanged just as loudly on the way in, accompanied by the whine of the winch.

That turned out to be only half the problem, however, as with my mind in turmoil, I couldn't read the measurements marked on the chain—and I'd painted them on myself. I then remembered changing the code halfway through, but still the markings didn't make any sense. Eventually, I recognized that Dave had attached the wrong end of the chain to the anchor.

Finally, the anchor was down in the right spot with twenty-five feet of chain out. Dave now slowly backed up the boat, while I let out another fifty feet. He then continued backing up, dragging the anchor across the bottom. I stood next to the windlass with my foot resting on the chain, expecting to feel a change in vibration when the anchor hooked. Nothing happened. We went through the entire process three more times, but still, nothing happened.

Some kind soul, undoubtedly thinking we were going to go on all night, then hollered, "Maybe you could try coming in from the side and pulling the stern around from shore!" Following this very logical piece of advice, we succeeded on the second try, as we were then using the length of the cove instead of the width, so had more room.

It was a rude awakening for two people who thought themselves reasonably competent. I was particularly humiliated and tried to redeem myself by taking the stern line ashore. Putting my canoe in the water, I paddled away with a feeling of relief, knowing this was something I could handle. As it turned out, pulling the stern of a fourteen-ton boat around ninety degrees was no easy task. I persevered, while the sweat poured off me in buckets, because there was no way I was *not* going to do it. However, I then spent the rest of the night wondering whether I was cut out to be a sailor.

My spirits didn't lift until morning, when I stood at the helm admiring the soft, clear light that bathed the quiet cove. After Dave had the anchor up, I steered *Windy Lady* out into the channel, where the mirrored surface of the water reflected a picture-perfect day. Twenty minutes later, just as the sun warmed my face, I felt a current of air. We were then approaching a pass that would take us from sheltered waters out into the Strait of Georgia (Day 2).

Once through the pass, I turned the bow into wind,

bringing the boat to a stop while Dave hauled the mainsail up the mast. I then pulled out the headsail while he went below to shut down the engine. Soon we were sailing northwest up the strait, tacking back and forth into 18-kt winds. With no reefs in either sail, I found the boat difficult to handle but he didn't seem to mind. In fact, he soon steered her farther out into the strait, where stronger winds sent the boat racing across the water.

With both sails filled, *Windy Lady* cut smoothly through the waves, full speed ahead. But suddenly, a strong gust of wind spun her around violently, knocking her down. Within seconds, she was dead in the water, deck perilously atilt, with the bottom edge of the headsail only inches above the waves. Dave was then struggling to hold the helm, but the boat only returned to an upright position when the 26-kt gust released its grip on the sails.

Alarmed, not knowing what had happened or why, I went below and checked the cabin. The violence of the incident was then evident; locker doors stood open, their contents dumped out onto the sole, cushions were scattered about, as were books and charts from the tables. Worse yet, seawater had come in both galley and head sinks, spilling over onto the sole, while fresh water had sloshed out of the main water tank.

Dave now remembered that we'd left the galley thru-hull open, so we started *Windy Lady* sailing and tacked to port, causing her to heel over on the opposite side. The sinks then drained and I closed the thru-hull. After quickly stuffing things back into lockers, I wiped up most of the water, but was then ready to call it a day. Dave was the captain, however, and he wanted more "experience".

We sailed for two more hours, crashing into choppy, five-foot waves that sent spray flying up over the foredeck, soaking the foot of the headsail. We now learned that the deck was dangerously slippery when wet, and there was nothing to hang onto forward of the mast. We finally started the engine and furled the headsail just before entering a narrow, rocky passage leading into a nearby anchorage.

Strong gusts buffeted the mainsail as we transited the passage, heeling the boat over and turning the bow, threatening to put us on the rocks. A final gust then hit straight on, shaking the sail and boom quickly from side to side, producing a loud, irritating clanging. The racket came from the boom, where we had stowed a long-handled brush. I was tired and the noise so grated on my nerves that, without thinking, I stepped over and grabbed at the brush, intending to remove it.

Just as I pulled the head free, the boom again jerked sideways, yanking it out of my hands and slinging it astern. Disbelievingly, I turned—and saw Dave. He was standing directly behind me with a dazed look in his eyes. Horrified, I watched as a lump appeared in the middle of his forehead where the round head of the brush had hit. I was appalled, knowing my impulsive action was responsible and have felt guilty ever since. Fortunately, he never even complained of a headache.

As the anchorage was large and exposed to the wind, we let out 125 feet of chain in order to reduce the strain on the anchor. Unlike the previous night, we were successful on the first pass. I then restored order to the cabin and scrubbed down counter and floors before supper. The wind kept *Windy Lady* dancing on the end of her mooring line all that night, with the occasional loud groan echoing underwater as the chain dragged across the bottom.

Come morning, winds were gusting from 20–30 kt, so we stayed put, which proved fortunate as the head plugged up. Dave, in a foul temper, then spent hours removing hoses and cleaning out the buildup of calcium that blocked them. When the sun came out and the winds dropped to 15 kt later that afternoon, we toured the foredeck and discussed changing the headsail.

The genoa that we were using was a huge sail, meant for light winds. We figured that the working jib, which was smaller and heavier, was better suited to existing conditions. I was surprised that Dave wanted to do the sail change then and there, as at the marina, I'd noticed large sails were difficult to handle even in light winds. But when I protested, he quickly ended the discussion by asking, "Are we going to sit around at sea and wait for light winds?"

Windy Lady's headsails were triangular and one side slid into a track running up the forestay. The track was part of the furler, which wrapped the sail around the stay when it wasn't in use. Various ropes controlled the sail: a halyard attached to the top pulled it up the track; the furling line wrapped it around the forestay; and two sheets were tied to the free corner (the clew), so that it could be pulled out to either side of the bow.

Throwing the free ends of the sheets onto the deck, Dave took a firm grip on the clew, pulled out a few feet of sail, and ordered me to release the furling line. When I did so, the wind caught the sail, instantly pulling half of it off the furler, ballooning it out over the bow, and ripping the clew from his hands. Although he managed to hang onto the sheets, the sail then whipped back and forth, dragging him all over the foredeck. As he couldn't let go, I tried re-furling the sail, which happened to work.

Not at all discouraged, he tied the clew to the mast, which kept the sail under control while we pulled it down. After rolling it up and hauling it below, we brought up the working jib. It was a new sail, so the material was heavy and stiff, and we had difficulty pulling it up the track, even with a winch. As we raised it higher, the wind began snapping the sail across the deck, causing the bow to slew about on the anchor chain.

We didn't pay much attention until we heard a sharp clanging noise and turned to see the sail snapping wickedly at head level, with the metal ring sewn into the clew striking the mast. Dave immediately darted forward, grabbed the sail midway along the foot, and began gathering it into his arms. Unfortunately, he couldn't hold enough of the sail to subdue the clew, which continued snapping about his head.

Scared to death that he would be struck at any moment, I managed to partially furl the sail. He then grabbed the clew, and we attached a sheet and secured it. In my relief, I was suddenly furious; it wasn't just that the danger was past but also that my earlier suggestion to tie a sheet to the clew had been ignored.

On Day 4, winds were lighter and the boat easier to handle, so we had a pleasant sail up to Nanaimo. We took our time at anchoring and then figured out the best way to winch our new fiberglass dinghy over the side. Lowering the 5-hp Johnson engine was more of a problem; I found it heavy and awkward to handle and struggled to pass it down to Dave. When it didn't run, he then spent the remaining daylight hours tearing it apart and half the next day putting it back together.

Late that afternoon, we dinghied ashore and found our way to a chandlery, where we purchased the lines we wanted before heading north. We then sat out in the cockpit with a piece of three-strand rope, splicing an eye in one end and back splicing the other end. We would use it as a snubbing line, meant to take the stress off the windlass when we were at anchor.

Next morning (Day 6), we put two reefs in the mainsail before leaving Nanaimo harbor and sailed very comfortably in 18–20 kt winds and quite wild seas. When the breeze died two hours later, we started the engine and steered out into the Strait of Georgia. The strait was eighteen nm wide at that point, and once we got farther out, the winds picked up to a steady 12–15 kt.

The sailing that afternoon more than made up for the difficulties we'd so far encountered. At first, I was simply grateful for the quiet when Dave shut down the engine, but soon *Windy Lady* was gliding smoothly through the water at five kt and I was enchanted. I stood at the helm, gazing out at a deep-blue sky and sparkling dark waters dotted with whitecaps. With my body swaying gently, I could feel sun and wind on my face and hear the soft rustle of waves washing against the hull. It was magic!

That night, we anchored easily in the muddy bottom of a small bay on an island in the middle of the strait. We were underway early the next morning (Day 7), expecting a long day to the next anchorage on the north side. With warm sunshine and steady winds of 15–20 kt, we spent the morning experimenting with the sails and learned all about our boat's "no go zone".

A sailboat cannot sail directly into wind. Some boats can steer about thirty degrees off the wind, but *Windy Lady* was closer to forty. If we wanted to steer a course of 315 degrees (NW) and winds were on the nose, we had to steer either 275 or 355 degrees (almost straight west or north). We would then tack, turning back and forth through the wind, in order to go in that direction. That day, we sailed thirty nm in six hours but made only fifteen nm towards our destination. The wind then died, leaving us out in the middle of the strait, and we motored for hours.

The deep water and shelf-like, rocky bottoms of the anchorages in the Copeland Islands presented new challenges, and we went on to a second cove before successfully anchoring. An eerie rumble then reverberated through the water as the boat swung with the tide, dragging the chain across the rocks. With warm, sunny weather, we stayed for three nights, taking the time to explore the marine environment.

I paddled around the bay in my canoe at low tide, noting that sea lettuce covered the rocky intertidal zone, while every crack in the rocks had a resident purple starfish. A long stream of water shooting through the air led me to a clam bed, where water was squirting in all directions. I glimpsed a small animal as it disappeared into the bush, a weasel I think, and saw a seal resting easily in the water near the bow of the sailboat.

Meanwhile, Dave went fishing, taking the dinghy around to the far side of the island. He saw a powerboat pulled up on shore, but didn't pay any attention until he heard whistling and turned to see a man waving his arms. When he motored over, the chap met him at the water's edge, saying, ""Am I glad to see you!"

With some prompting, the fellow explained, "I left Lund about 1900 last night, that's a small village over on the mainland. I was going to a fish hatchery about forty-five minutes up the coast. But my engine quit, so I drifted around all night, then washed up on shore about 0500. I have no idea where I am."

Dave was happy to help, explaining where he was and giving him a hand hauling the boat down to the water. After watching the fellow yank on the starter cord, he then asked, "Could you be out of fuel?" The castaway, who still smelled strongly of liquor, switched over to a second tank; moments later, the engine caught and he was off in a cloud of spray.

After leaving the Copeland Islands, we spent three days threading our way through the maze of channels between Vancouver Island and the mainland. The sunny weather continued and we sailed when we could but mostly used the engine. We grew more confident with every hour that passed as we raised and lowered sails, motored through narrow passages, studied GPS, radar and engine instruments, charged batteries, monitored energy use, and conserved our limited supply of water.

On Day 10, with no wind, we motored north into Desolation Sound. I wondered why it was that Captain Vancouver had saddled the place with such a bleak name some two hundred years earlier. Looking around, I noted the deep, dark waters and high mountains, many of them with steep, rocky faces. It felt closed in, more like a lake than an arm of the sea.

Late that afternoon, we cautiously entered a marine park and anchored near a small stream rushing down over the rocks at one end. In the hours that followed, the air grew still and the water calmed, becoming a mirror that perfectly reflected the surrounding mountains and forest. Captivated by our surroundings, we set out to explore the cove. In the process, we learned that the dinghy was rather small for the two of us, and the 5-hp Johnson motor barely adequate.

Strong outflow winds funneled down through a notch in the mountains when we anchored in a small, isolated bay on Day 11. Thinking that conditions might well serve as our final exam, we took extra care in setting the anchor. I guess we passed, as we were still there come morning.

Day 12 brought us to Yaculta Rapids, a mile-long passage with tidal currents of 5–7 kt. Nervously, I checked the tide table repeatedly because we needed to be in position at the south end of the passage before slack tide. A fish boat was waiting when we arrived and started up the channel about fifteen minutes early, according to my reckoning. Conceding to local knowledge, we followed it up a back eddy on the eastern shore.

The current caught *Windy Lady* as soon as she poked her nose out into the main channel, turning her almost broadside before she started to come around. The bow then slewed from side to side as she slowly pushed up through the lower rapid. When the channel turned sharply, we couldn't find the marker buoy until we were almost on top of it. The boat slowed noticeably through the turn, and with my stomach in knots, I kept a wary eye on nearby whirlpools. After crossing a wide bay, we proceeded through the upper rapids, at which point we were free and clear, another obstacle overcome.

Continuing under power, we now relaxed and studied our surroundings. During the next few hours, we saw dozens of bald eagles; the big birds, both adults and juveniles, flew low over the water, perched on the rocks, and fought amongst themselves. We anchored that night in another isolated bay, where a pair of mergansers watched from the safety of the reeds, while a small hummingbird checked out the cockpit.

Late that evening, with the sun low in the sky, we heard the throb of a diesel engine as a fish boat entered the bay. It circled round a small log boom tethered nearby, jumping and puffing as it worked. Before long, the skipper had picked up the boom and the boat was heading out. As silence returned to the bay, the songs of hermit thrush drifted across the water. Then, in the deepening twilight, the wild cry of a loon split the air.

Awakening to the sound of the chain dragging across the sea bottom, we were underway by 0530, greeting the sun as it peeped over the hills (Day 13). With calm air, rippled seas, and an ebbing tide, we made good time and were soon turning west into the narrow, deep waters of Johnstone Strait. This was the main route along this section of Vancouver Island and we now left the maze of channels behind.

Before long, we could see whitewater ahead, as outflow winds whipped down an intersecting channel. *Windy Lady* was soon tossing about in rough five-foot waves pushed up by 20-kt winds. When Dave saw the dinghy bucking and twisting at the end of the towline, he felt compelled to rescue the outboard motor. I was appalled and tried to talk him out of it, but he didn't hear a word. Methodically, he set about his task: reducing engine power, adjusting the autopilot, pulling the dinghy up close behind.

With the boat lurching about in breaking waves, he climbed over the transom and stepped down three feet onto the narrow bow of the dinghy. Sliding his way to the back, he loosened the clamps on the outboard, lifted it up and pulled it inwards. With all the weight at the back, I was sure that both he and the engine would go overboard, but they didn't. He somehow manhandled it into the dinghy, dragged it forward, and then stood and kept his balance long enough to hand it up to me.

Dave took his physical strength for granted, but I

knew my limitations well. I was almost a foot shorter and ninety pounds lighter; I couldn't lift the engine up into the cockpit at the best of times. Knowing from the start that I would have to be part of this foolhardy venture, I seethed with rage. But I also knew that if I dropped it, I might as well follow it into the sea. So, without pausing to think, I leaned out, draping my upper body over the transom, then grabbed hold and gave a mighty heave—and swung the engine up into the cockpit.

We were through the worst of the waves in about fifteen minutes, but ran into similar winds at the next two junctions, although seas were never as bad. The winds grew lighter as the afternoon progressed, and with a favorable tide, we kept going. After motoring forty nm, we anchored late that afternoon on an island on the north side of Johnstone Strait.

I stayed in a foul mood for hours and was still furious when I wrote in my journal that night. I complained that Dave was putting himself, *Windy Lady*, and me in jeopardy. It's obvious that I deeply resented being forced to participate in this latest escapade. It's also apparent that I was rubbing up against the boundaries of my comfort zone, had been doing so for days, and really didn't like it. What I didn't understand then was that those boundaries, both mental and physical, were self-imposed and far too comfortable. Learning to stretch them was the most important lesson of the shakedown cruise; ultimately, that was what most prepared me for going to sea.

CHAPTER 3

A Different World

The next morning was clear and calm, with the mirrored surface of the bay reflecting a perfect image of the surrounding forest. The sun was already warm when we raised the anchor at 0615 (Day 14), and captivated by the view, I stayed at the bow while Dave steered the boat out and around the point. Air and water were then so still, we seemed to glide into a picture with forested islands floating on tranquil waters and snow topping the mountain ridges.

I soon noticed a small patch of fog nestled against the south shore of Johnstone Strait; within a mile, it had spread across the entire channel. With sky and shoreline disappearing, Dave ordered me to the helm, while he went below to the chart table. Falling back on skills honed in his flying days, he plotted our GPS position on a chart and established waypoints that would keep us safely in mid-channel. Monitoring our progress, he then called up instructions as needed to keep *Windy Lady* safe and on course. As well, he kept an eye on the radar screen, watching for traffic and noting the contour of the shoreline.

Meanwhile, I stood at the helm, keeping the bow pinned to the compass heading that he had been steering, and watched as dense fog closed in around the boat. Soon I could see only ten, maybe fifteen feet into a dull, grey mist. When ordered to do so, I changed course blindly, steering into nothingness. Before long, I had an eerie sense of being isolated in time and space, as if a cocoon encasing the boat was carrying us through a silent, ghostly world. My only links to reality were the dim light shining up through the companionway and the throb of the engine.

Sometime later, I heard a faint noise and turned to see maybe a dozen dolphins swimming near the port quarter. I excitedly called to Dave, and we watched them swim in close on the port side, dive, and then surface to ride the starboard bow wave. Seconds later, they dropped back and disappeared into the greyness, but their images remained in my mind. They had been solid objects in a formless world.

About midday, Dave called up a warning; we were approaching the narrow channel between two islands. Soon, I was searching the greyness for shadows that might reveal their locations, but saw nothing. I did notice a log, mostly submerged, drifting slowly past and recalled that we had timed our departure to arrive there at slack tide. I looked around for more debris but couldn't see the water ahead, and then heard a few thumps as unknown objects bumped against the bow.

Unseeing and unseen, we motored west for six long hours, covering thirty nm as we passed down Johnstone Strait and out into Queen Charlotte Strait. By 1300, the fog wasn't quite as thick, and we could see some fifty feet around the boat. I was still on watch but no longer stood at the helm, as Dave had turned on the autopilot. When the fog showed no sign of lifting, we decided to head for Port Hardy on the north end of Vancouver Island; it was about twelve nm to the west and we figured would be the easiest harbor to enter.

Half an hour later, the sky started to clear, and within minutes, the mountains on the island appeared. A light wind came up as the fog dissipated, so we happily reverted to our original plan, raising the sails and steering northwest, to an anchorage on the far side of the strait. Our enjoyment was short-lived though, as we were soon pushing into 20-kt headwinds and lumpy, three-foot seas. We then found ourselves heading toward a small, rocky island in the middle of the strait, so had to tack.

I was at the helm but have no idea what happened next. Maybe my timing was off, or maybe a wave or gust of wind caught the bow. All I know for sure is that one moment the boat was speeding forward; the next, it was dead in the water and the deck was heaving and twisting beneath my feet. In fact, the waves had turned *Windy Lady* broadside and were tossing her about like a toy. Both sails were backwinded, and the wind was whipping the headsail, causing the sheets to thrash the water viciously. Dave jumped to the winch and furled it, and we then tried to start sailing with the mainsail. When that didn't work, he started the engine.

That was the beginning of a three-hour-long ordeal, as *Windy Lady* plunged into high, steep waves pushed up by gale force winds of 28–32 kt. Waves broke over the bow, sending spray flying over the foredeck, while water ran down the cabin roof and under the windscreen into the cockpit. Occasionally, a bigger series of waves sent the deck pitching ever steeper, and then Dave, who was at the helm, was on the receiving end of the spray. Inside the cabin, settee cushions went flying, locker doors popped open, and seawater ran into the cabin through the air vents.

As we drew closer to the north side, we anxiously peered at the misty shoreline, looking for small islands that hid the harbor entrance. When we eventually spotted them, Dave turned the bow in their direction, wanting to find shelter and drop the mainsail. Almost immediately, two alarms began to shriek, filling the small space inside the cabin. Knowing instinctively that the warnings came from the depth gauge, he jerked the helm back. He then tried repeatedly to steer towards the islands, but each time the shrill alarms screeched unbearably. Finally, however, he found a safe route. We then dropped the sail and located the harbor entrance with the help of the GPS. Soon we were safe and secure inside, able to relax for the first time in eleven hours.

Two large power yachts arrived in the harbor just before dark and crept out at 0500, in a cool, grey dawn. Dave heard the purr of engines as they left and, bothered by their early departure, tuned in the weather broadcast on HF radio. The area forecast for the open coast to the north was for low swell, light wind, and visibility of less than a mile in fog. It was a good forecast, if we wanted to continue north around Cape Caution.

The current plan was to head west, back across Queen Charlotte Strait to the north end of Vancouver Island. However, we had toyed with the idea of continuing up the Inside Passage to Ocean Falls, which would add 250 nm to our route. Now, with weather cooperating, we couldn't resist the temptation. At 0600 on Day 15, we motored out into the calm waters of the strait and turned the bow northward. Off to the southwest, the looming shape of a large fogbank rose high in the sky, the only visible reminder of the previous day's drama.

Windy Lady was soon rolling in a slight swell pushing down from Queen Charlotte Sound; as the wide, shallow trough deepened, the roll became more pronounced. The air was misty at Miles Inlet, with clouds sitting low on the hills, and we checked the weather report one more time. We'd been apprehensive about traveling along this stretch of open coast, but rounding Cape Caution proved a non-event. After motoring for five hours, we simply passed by about a mile offshore, noting only that the boat rolled heavily in water less than 100 feet deep.

An hour later, winds were gusting from 12–17 kt and seas were building. Both subsided as we turned into Smith Sound, where we found our way into a quiet cove. We stayed over a day, and I used the time to study charts and guides covering the area to the north. Dave went fishing, disappearing with the dinghy for hours, and returned with several rockfish. Of course, as soon as he was out of sight, I started thinking about how isolated we were and drove myself crazy with all sorts of imaginings.

That night, swells pushed up into the cove, causing the boat to roll uncomfortably. When the early forecast then called for increasing winds and a low-moderate southwest swell, we decided it was time to move on (Day 17). We set off under power with Dave at the helm, and at the entrance to the sound, I was left staring in round-eyed wonder as hills and valleys of moving water swept around the boat.

We continued seaward until we were clear of the peninsula to the north, then Dave turned the bow and followed the coastline. Soon we were crossing the wide entrance to Rivers Inlet, and for the next three hours, swells sweeping beneath the hull rolled *Windy Lady* far over onto her sides. Dragging at the stern, they turned the bow and left her wallowing in the troughs, so we spelled each other at the helm as we fought to keep her moving north.

The waters gradually calmed as we proceeded up Fitz Hugh Sound, in the lee of Calvert Island. This well-traveled portion of the Inside Passage was busy, and we saw three multi-storied cruise boats, a BC Ferry and numerous tugs, barges, and powerboats. We also saw a huge pod of dolphins, with hundreds of animals leaping down the channel toward us. Half a dozen arrived to check our boat, circling around and diving beneath it. With grinning mouths and boundless energy, they seemed to embody the joy of living, and I felt a warm glow long after they'd disappeared.

After a long day motoring, we anchored in a quiet lagoon off Fisher Channel, and that night saw our first rain. By morning, clouds were sitting low on the surrounding hills and the air was cool. Our six neighbors were all underway by 1000, leaving us to enjoy the quiet solitude on our own. Two boats then arrived late in the afternoon, traveling together, and anchored at the far end.

The sun began streaming through breaks in the clouds as it dropped in the west, and the air grew still; soon the water was a mirror, reflecting the surrounding old-growth cedar forest. I noticed a great blue heron standing motionless on a nearby float, and then heard the irresistible cry of a loon pealing over the treetops. Climbing into the dinghy, we set off down the cove, with Dave quietly rowing while I strung a fishing line out behind. Three seals immediately took up position around us, making it clear that if we caught a fish, we wouldn't keep it.

As I listened to bird songs floating across the water, something in the gloom on shore caught my eye. Nothing moved for so long that I decided I was mistaken—and then a small deer flicked its tail. Dave let the dinghy drift closer, and at seventy-five feet, the animal turned and walked away, then bolted into the forest. As we returned to *Windy Lady* in the deepening twilight, a pair of Canada geese flew overhead, followed by two small ducks. The magic continued even after we were back onboard, as I noticed dimples in the dark waters nearby. Peering closer, I saw flashes of silver as a school of herring circled below.

Next morning (Day 19), an otter swam casually around the end of the float when we raised the anchor at 0630. With no wind and calm waters, we then motored west under a grey, overcast sky. The clouds grew heavier and a light mist was falling when the small coastal village of Bella Bella came into view three hours later. A huge Holland-America cruise ship then blocked the channel in front of us, dwarfing the buildings and the island, and we pulled over to the side and waited while it passed.

Rain was falling steadily when we tied up at a dock at the nearby marina. We found our way to the store, where we bought two charts and a dozen eggs, and chatted briefly with a couple on their way to the Queen Charlotte Islands. Joining some two dozen customers in the café, we then ate greasy hamburgers and breathed in a lot of second-hand cigarette smoke. The din in the room drowned out the TV news but we did hear one story, as everyone stopped talking long enough to listen to a report about a large sailing vessel that struck a reef and sank.

At high tide that afternoon, we motored through the narrow, rocky channel known as Gunboat Passage. Channel markers were hard to spot, as visibility was poor in rain, so I worried about the keel all the way. We then stopped at a nearby anchorage, and later that evening, Dave pointed to some very large rocks blocking the entrance and asked, "Did we come in past them?"

With a twelve-foot tide, there had been no sign of them earlier, but I was the navigator and should have known, so confessed, "Yes, and I didn't know they were there!"

The next day was again grey and cloudy, not particularly good for travel, so we stayed over. I paddled about for five hours in my canoe, and checked out a deep-water route by the rocks at the entrance. Clouds sat low on the mountainside when we left the following morning, and although the tide was low, the depth gauge never read less than twenty-four feet (Day 21).

We soon turned north into Fisher Channel, where thousands of small ducks floated in a long line down the middle of the passage. I checked with binoculars and identified a few Bufflehead, but most were Common Golden Eye. When we rounded the next point, we spotted about two dozen seals lying on the rocks.

By midday, we were in the isolated community of Ocean Falls, tied up at the dock in 100 feet of water. We talked to a few locals in a small nearby café and learned that the paper mill had shut down in 1982, after operating for some seventy years. The town's population was back up to 160, after dropping to a low of 25. The afternoon now turned sunny and warm, so we spent three hours wandering about the old town site, dam, and lake. High on a hill, hidden in an overgrown street, we found a plaque that read, "Owen, Jack and Bert lived here for 46 years, to 1989", and then, "Home Again".

Flowers of all kinds bloomed on the hillside, with domestic plants and native species intermingled. Yellow day lilies and purple irises grew alongside buttercups, daisies, and columbine, with cow parsnip, devil's club and thimbleberry thrown in. But only when I noticed the little blue forget-me-nots did I think of the people who had lived there, of the hopes and dreams long forgotten and buried beneath the undergrowth.

Although we enjoyed visiting these remote, coastal spaces, we knew it was time to return south. We weren't gaining any sailing experience in these long, narrow inlets. We waited a day though, as a storm with 50-kt winds and twenty-foot seas was approaching the west coast. Rain fell overnight with a few showers the following morning, and then the sky cleared.

We started south on the afternoon of Day 23; again, we had no wind and calm seas. As we approached the site of the old cannery in Fitz Hugh Sound, I saw a flock of sea gulls circling in the sky and splashes of white dotting the water below. The splashes proved to be dolphins, and soon sunlight was glinting off dorsal fins. Only a few of the animals were jumping, but I watched delightedly when one tail-walked four times. The next morning, I counted as one tail-walk thirteen times in a row!

After motoring for four hours, we passed by the deserted cannery buildings and entered the anchorage just before high tide. Dave circled the boat around, checking for depth, then picked a spot and brought *Windy Lady* to a stop. I dropped the anchor off the bow, realizing then that the process was no longer a leap into the unknown. I now knew how much chain to let out and, with my foot, could interpret the vibrations, so had a good idea of what was happening below. Feeling good, I gazed around the inlet while we set the anchor. The sun was dropping in the west and shadows were lengthening, but that didn't hide the sobering sight of an oil slick on the water and rusty barrels strewn about on shore.

When we left the anchorage early on Day 24, winds were light but swells were pushing up the channel. We then stopped to raise the mainsail, and just as Dave heaved on the halyard, a large swell rocked the boat. Instantly, the line wrapped itself around two of the upper mast steps. Climbing the mast was a job that I had claimed, partly because Dave seemed responsible for everything else. I usually didn't mind but that day another large swell hit when I was twenty feet above the deck. With the mast swaying out over the water on one side, then the other, I just closed my eyes and hung on. As soon as the rocking eased, I quickly freed the halyard and scooted back down.

During the next few hours, we sailed and motored, then sailed and motored again. Halfway down Calvert Island, the breeze picked up to 20 kt, and soon swells were pushing up from Queen Charlotte Sound. Before long, the waves were looking pretty high, probably close to ten feet from trough to crest. As *Windy Lady* was bobbing around like a top, we once again spelled each other at the helm.

We anchored that night in Rivers Inlet and awoke to clear skies and an early morning chill. Enjoying the solitude, we lingered in the cockpit with our coffee. When the noisy squawking of a stellar jay disrupted the silence, we traced the sound to flashes of blue diving at an eagle perched in a nearby tree. The jay then hopped from branch to branch, scolding stridently, and the eagle finally moved to another tree down the bay. That didn't satisfy the jay and it followed, bringing a couple of friends. After a little more haranguing, the eagle gave up and flew away.

We left the anchorage at 1030 on Day 25, setting off under sail in a light SW breeze and calm seas. The wind died away soon after, then promptly picked up from the NW and quickly grew stronger. By the time we reached Major Brown Rock at the entrance to the inlet, winds were nearing 20 kt and seas were growing. *Windy Lady* was now running on a beam reach and grew increasingly difficult to handle as she tossed and turned. When a couple of strong gusts each pulled the bow around fifteen to twenty degrees, I'd had enough. Nervously turning to Dave, I pleaded, "Don't you think maybe it's time to put a reef in the mainsail?"

Gesturing with his hand towards the twisting, heaving foredeck, he retorted, "If you think I'm going up there, you're crazy!"

He took the helm and I quickly realized that he really wanted to see how fast the boat would go. I grew more anxious because seas were wild, with the ebb current from Rivers Inlet and northwest chop from Fitz Hugh Sound mixing it up with a strong westerly swell. We continued to head out to sea for another half hour, during which time the boat speed went from seven kt to nine. Meanwhile, 25-kt winds had heeled the boat over, so that waves covered the lower half of the portholes.

Finally satisfied, he turned the bow southward, following the coast around to Smith Sound. Within minutes, an even stronger gust hit *Windy Lady,* violently spinning her around ninety degrees and laying her over onto her side. With that, he raced below and started the engine. I furled in the headsail, knowing that whether we wanted to or not, it was time to reef the main.

Turning the bow into wind now put us at odds with the sea, and spray flew everywhere as the foredeck heaved and twisted violently. I let the bow fall away while Dave prepared the reefing lines, but had to turn back when he started working with the sail. When I did so, the deck again heaved, rolled, and turned, all at the same time, seemingly intent on knocking him off his feet. At one point, a huge wave rose up in front of the bow and I tried to yell a warning, but my throat was so dry that I had trouble getting out the words. Still, he heard and grabbed on with both hands, but the boat just rode up over top of it. It seemed a long time before he was once again safely in the cockpit.

With the mainsail reefed, we motored south, but strong seas sweeping under the keel made it difficult to keep the boat on course. We fought the helm for nearly two hours and then turned into Smith Sound. Still, the struggle wasn't over, as swells pushing on the stern attempted to turn the boat broadside, first to one side, then the other. The seas only began to subside as we crossed behind Brown Island, and 20-kt gusts kept the boat dancing even in the anchorage.

With clear skies and warm sunshine, we relaxed the next day; I paddled about in my canoe for hours, while Dave did more fishing. The early forecast the following morning wasn't very promising. A low-pressure system developing to the west of Vancouver Island promised to send a series of storms onto the west coast. Deciding that we should get back around Cape Caution, we quickly raised the anchor.

We left Smith Sound on Day 27, motoring out on an ebb tide strengthened by melt water from the mountains. At the entrance, we met a moderate westerly swell, and *Windy Lady* rolled far over onto her sides, as we made the turn southward. I grabbed onto the binnacle for support, but heard Dave cursing from below where he was preparing breakfast. Swells hit us broadside all the way down to Cape Caution, and with light winds, the boat rolled heavily. Fighting to keep the bow on course, we each took a turn at the helm and ended up having to hang on a few more times.

Clouds sat low on the hills south of the cape, but the sky looked brighter to the southwest, towards Port Hardy. That proved an illusion, as we ran into fog near the western end of Queen Charlotte Strait. Dave again went below and took up his post at the chart table, while I stayed at the helm and watched as a group of islands disappeared in front of us. Piloting the boat through a grey, cotton-wool world wasn't as unnerving the second time around, but that otherworldly feeling did return, especially when the mournful wail of a foghorn drifted across the silent waters.

We were in the fog for more than three hours, and at one point, Dave hollered up, "Turn five degrees to starboard!" When I didn't respond quickly enough, he urgently added, "There's a boat dead ahead and there's three blips on the screen. I think it's a tug!" A short time later, I heard the throb of a diesel engine as the vessel passed by 300 yards to port; a few minutes afterward, *Windy Lady* rocked gently in its wake.

Winds finally picked up to 7–12 kt about twenty minutes after we made our last course correction. The outlines of mountains on Vancouver Island then appeared, followed abruptly by the looming shape of a small island to starboard. Minutes later we were in bright sunshine, with the sea sparkling deep blue and Hardy Bay only two nm away.

We cautiously found our way into a marina at Port Hardy, which was no bargain at $42/night, particularly with rotting boards and dog feces on the dock, and dead fish floating in the water. We then walked into town and treated ourselves to supper, but I felt exhausted as soon as I began to relax. Next morning, fog blanketed the harbor; it lifted, returned, lifted again, and then left a fog bank out in the channel.

We now phoned Environment Canada for a weather briefing and were told that the current low-pressure system would weaken in about two days' time. Northwest winds would then replace the existing southeasterly flow, with a brief weather window available between the systems in which we could leave. As it was apparent that we wouldn't be sailing offshore, we went shopping for charts, as ours didn't show the detail necessary for day hopping down the west coast. We didn't find even one.

We left Port Hardy early that afternoon, setting off under sail for an all-weather anchorage on the north end of Vancouver Island. Within an hour, winds were on the nose at 20 kt, so Dave started the engine. Winds and seas had calmed by evening, when we approached the impressive cliffs guarding the entrance to Bull Harbor. We took extra care in setting the anchor, and then sat out in the cockpit enjoying the solitude. The air was warm and still, with dark clouds overhead, and a continuous chorus of bird songs drifted across the water.

We spent three nights in Bull Harbor, with overcast skies, frequent rain showers, and a day of gusty winds. I paddled about in my canoe and we made daily excursions ashore. At the head of the bay, we came upon several abandoned houses, boarded up and in need of paint. Although overgrown by brush, Dave found a sign identifying the site as an Indian reserve.

Crossing the narrow isthmus to Roller Bay, we stood and looked out on open water extending all the way to Japan. Watching the surf pound against the steep, gravel beach, I wondered how it was that two such different cultures had developed on either side of that body of water. And as I stood there, looking out to sea, I again felt that overwhelming certainty that my future was out there.

CHAPTER 4

Expanding Our Horizons

On our last night in Bull Harbor, we went through the boat and carefully stowed everything we'd been using, then tied dinghy and canoe to the foredeck. Early the following morning (Day 31), we tuned in the HF weather broadcast and learned that the low-pressure system had stalled some 300 nm to the west. A second low was now strengthening off the Oregon coast, so there would be no weather window. We had a choice of leaving immediately, or maybe not going at all.

I followed Dave up on deck and watched as he studied the grey, overcast sky. Winds were light on this side of the island, but there would be 20-30 kt headwinds and strong seas on the thirty-nm run down to Winter Harbor. Quickly making up his mind, he turned to me and said, "We're going. Let's get the anchor up."

Trying to hide my apprehension, I took a last lingering

look around the snug harbor as I walked up to the bow. We had no time to waste, however, as we wanted to cross the shallow bar off the end of Vancouver Island at slack tide. With light winds and calm seas, we rounded the end of the island and crossed the four-nm-wide bar with no difficulty. We then ran into swells when we were still several miles from Cape Scott on the northwest tip of the island; they grew bigger as we drew closer to Scott Channel.

Four and one-half hours after leaving Bull Harbor, we stopped on the verge of a stormy-looking sea. We raised the mainsail, putting in two reefs, then pulled out the headsail and shut down the engine. Dave took the helm and steered the bow out to sea, heading into SE winds of 20 kt; ninety minutes later, we were seven and one-half nm from Cape Scott. We then tacked back towards shore for forty-five minutes and, unbelievably, were only five nm from the cape.

Winds increased to 30 kt soon after, so we put a reef in the headsail and continued to tack back and forth. With headwinds and a strong southwest swell, maybe even some current, we fought for every additional mile. Sitting in the cockpit, sheltered only by the windscreen, I watched as Windy Lady crashed into seas that grew ever rougher. Bobbing and twisting confidently, with spray flying high over the foredeck, she had no problem with the conditions, and my trust in her was growing.

By 1700, we'd sail twenty-one nm but made good only seven and one-half. We decided then to head for a storm anchorage about two nm away and started the engine. Not five minutes later, a squall knocked the boat onto its side; waves broke over the bow, washing back over the cabin roof, and others broke alongside, dousing the cockpit. *Windy Lady*, however, instantly bobbed upright, shook herself off, and struggled onwards.

During the hour it took us to reach the anchorage, I sat huddled in the cockpit with my stomach in knots. Frequent squalls brought strong gusts and driving rain, and at times, we seemed barely to inch forward. Waves crashed over the bow, sending the sea streaming across the cabin roof and side decks, with spray flying everywhere. But my heart leapt into my throat only once, when a following sea tried to turn the boat sideways as we swept through the narrow entrance into Sea Otter Cove.

Dave now proceeded cautiously, squinting through wind-driven spray as he tried to locate the mooring buoys. When he turned the bow towards the closest one, *Windy Lady* promptly ran aground on a rocky shelf. He easily backed her off, turned and followed the circular deep-water route shown on our chart. At that point, I walked up to the bow, dreading what was to come, because somehow I had to secure the boat to a buoy.

Kneeling down beside the toe rail, some six feet above the water, I coiled the mooring line in my hand, placed the boat hook nearby, and waited. The foredeck blocked Dave's view of the buoy, so he was steering blindly, but slowly we drew closer. When I leaned out on the port side, preparing to throw my line, I saw that the big metal storm buoy was mostly submerged. A squall hit us then, bringing a shower of cold rain and blowing the bow off to the side.

Undaunted, he repositioned *Windy Lady* and brought her up to the buoy once more. I threw my line down under the heavy bar that ran across its top, then reached down with the boat hook, pulled the rope through and brought the end back up on deck. As I secured it, I was ecstatic; I was dripping wet, windblown, and cold, but something had finally gone right.

Gusty winds jerked the boat against the short mooring line for much of the night, but eased just before daybreak. To our surprise, the five boats that had shared the anchorage were gone by 0600. When we left thirty minutes later, the depth in the entrance was under ten feet and low tide still two hours away.

With a light SE wind and smooth seas, Day 32 bore no resemblance to the previous afternoon. Relaxing in the cockpit, we had nothing to do but enjoy the scenery as we motored down to Quatsino Sound. We rode the swells in through the entrance, passing near surf breaking on rocks below the lighthouse, and tied up at the dock in Winter Harbor at midday.

We now renewed our acquaintance with a couple we'd first met in Port Hardy; they were on a sailboat called *Starkindred*. They'd had a rough crossing to Sea Otter Cove and then were towed into port when algae in the fuel killed their engine. We also met the American couples on two Nordic tugs that had shared our last two anchorages.

Lured out by warm sunshine, we strolled for an hour

along the shady tree-lined boardwalk that followed the edge of the bay. But when the store opened at 1500, we were waiting to search through their charts. Again, we found nothing. Gloria, off one of the Nordic tugs, then came to our rescue, lending us a coastal cruising guide. I spent an hour furiously reading and making notes, but there were many anchorages and I didn't know where we might stop.

The following morning brought bright sunshine, with strong winds that howled through the rigging of the boats. With the forecast calling for NW winds of 30–35 knots and moderate seas, we spent the day visiting and doing chores. I re-borrowed Gloria's cruising guide, expanding on the notes I'd previously made. Later, when I was mixing up a batch of bread, she came knocking on the hull, carrying a book that she wanted to show me. The title said it all, "Cape Horn: One Man's Dream, One Woman's Nightmare."

The 0500 weather report on Day 34 included a gale warning; it was gone an hour later, so we quickly prepared the boat for sea. A few dark clouds sat low on the mountains when we pulled away from the dock, but winds were light. Just outside the harbor entrance, I heard a sudden whoosh of air and turned to glimpse a grey whale feeding near the shore; it disappeared then resurfaced barely seventy-five feet off the stern. The whale looked to be about as long as our boat and its skin was rough, as though covered with barnacles. Mist rose from the blowhole as it again exhaled and disappeared.

We ran into moderate ocean swells at the entrance

to the sound, and as we turned southward, *Windy Lady* rolled far over onto her sides. I had to hang onto the binnacle to keep from falling across the cockpit, which made it hard to control the helm. So, I tried bending my knees to offset the rolling of the boat and instantly was able to stand erect. Thereafter, whenever the boat rolled excessively, I could be seen swaying in this very strange dance with the rhythm of the ocean.

Once we were clear of the entrance, we stopped and raised the mainsail but continued under power as we headed south around Brooks Peninsula. Since we'd started monitoring the weather, the reporting station there had consistently reported 40–50 kt winds, so we felt fortunate to have light winds and a sunny sky. Even so, small mountains of water surged around the boat as we passed to seaward of Solander Island. Our feelings of good fortune were further heightened by the sight of two large Orcas (killer whales) passing by some 100 yards out to sea, their tall dorsal fins rising six feet above the waves.

My confidence deserted me, however, when I saw the rocky outcroppings and barrier islands littering the approach to Walter's Cove at Kyoquot Inlet. Suddenly, my mud maps and handwritten notes seemed entirely inadequate. We had only one reference book onboard and it was no help, stating that local knowledge was required due to the many above and below-water rocks and reefs in the bay. So, I directed Dave to a round-about, deep-water route that took us to the start of a dredged channel leading into the cove. Trouble was, a sailboat with local knowledge crossed in front of us when we were halfway round, which didn't make the captain too happy.

I stood lookout at the bow as we entered the dredged channel, but then couldn't locate the next marker buoy. By the time I did, I could practically feel *Windy Lady's* bottom scraping on the rocks. My mud map then had some value, as it showed the relative positions of subsequent buoys, including the ninety-degree turns. We went directly to the store after tying up, but again found not one chart.

Dark storm clouds brought occasional heavy rain that afternoon, with SE winds of 25 kt reported offshore. With the boat safe and secure at the dock, we walked across the island to a viewpoint overlooking the ocean. As we strolled back around the harbor, I studied the two clusters of buildings sitting across from one another; both looked rather depressing on such a gloomy day.

Continuing down the road, we came to a small house, where a man was walking across the yard. He called out a greeting and soon was explaining that he was a teacher turned sawmill-operator. He seemed anxious to show Dave his band-saw mill, so I was taken up to the house and pawned off on his wife. She was sitting in a rocking chair, reading peacefully, and somewhat reluctantly marked her page, put down her book, and politely offered me a cup of tea.

As we chatted, the woman told me about her seven children, aged four to seventeen years, and her home, which was a converted one-room schoolhouse. I figured she had to be at least ten years younger than I was. Looking around the interior of her home, I noted the bare wood floors and walls, the simple furnishings. Two ladders at either end led up to a narrow loft that ran down one side of the room and contained what looked like large storage cupboards; I suspected at least two of the older boys slept there.

What followed was the strangest experience of my life, and I have no idea how long it lasted. But as we sat and talked, I started to feel uneasy, then grew confused. I tried to stay focused but something was wrong, and it dawned on me that I'd heard this conversation before, like many times before. To my horror, I then realized that I knew what this woman was going to say before she said it. In fact, I knew her; I knew everything about her.

In some strange way, past and present merged and I was remembering a time when I was a child, and my family had lived in even simpler conditions in bush camps in the BC interior. As I struggled to understand what was happening, my sense of connection was so powerful that I asked myself, "Is she living an alternate version of my life?"

With my mind flitting back and forth, I now saw the image of a wood stove in a small kitchen, but I didn't recognize it. Becoming completely rattled, the question then became, "Whose life is this, hers or mine?" I even remembered thinking, "But I'm going to sail across the ocean!"

I must have made a very strange visitor and when the men returned, I quickly dragged Dave away. Only when we were back on the boat did I start to feel reassured. I knew I had been thinking of my mom, who raised eight children in somewhat similar circumstances. What really threw me though was the notion that when I was born, friends and family would have predicted exactly such a lifestyle for me.

In truth, my life had turned out very differently from my mom's, and long ago, I had credited that fact to the time and place in which I lived. Now, such a pat answer no longer seemed adequate. I could plainly see that my life was a continuation of the lives of my parents and grandparents. I was only in a position to take advantage of a changing world because their decisions, their struggles, had placed me there.

We left Walter's Cove on the morning of Day 36, motoring out through the dredged channel at high tide. With a ten-foot swell in the bay, we wrapped two halyards around the upper mast steps, so I again climbed a swaying mast. With light winds and clear, sunny skies, we motored south, enjoying impressive views of rugged mountains rising up out of the sea. Unfortunately, the movement of the boat, rising and falling in high swells, magnified the visual impact of scars left on steep slopes by slides and zigzagging logging roads.

Shortly before midday, I noticed the shiny black backs and small fins of two Dahl dolphins swimming about fifty feet off the bow. Soon after, we noticed fish jumping and Dave hurried below for his fishing rod. Ten minutes later, he sat beaming from ear-to-ear, admiring the four-pound Coho salmon he'd landed. We continued on past Esperanza Inlet, where heavy surf pounded the rocks, spray flew high in the air, and a misty cloud hung over the shoreline.

I best remember Nootka Sound, however, as there the sound of the engine changed and it died. Rushing below, Dave raised the sole, tied it back, and started looking for the problem. I reset the sails but *Windy Lady* made only two kt in the light winds, providing no steerage. Looking out on the huge swells sweeping around the boat, I realized we were adrift. I then sat and stared at the swells, the surf, the isolated mountains on shore, and began to feel uneasy. Soon, I was worrying about washing up on the rocks and wondering whether there were any boats nearby that could rescue us; unconsciously, I tried to steer towards the inlet.

Abruptly, I realized that I was being absurd. We were a sailboat! We had to be able to manoeuver without an engine! The real irony struck me moments later. Here I was, planning to sail thousands of miles across an ocean, yet obviously doubted our ability to deal with a minor emergency in light winds on a sunny afternoon. With that thought, I called down to the captain requesting permission to tack, and then turned the boat back on course.

While my actions were of little consequence that day, it was another important milestone. For the first time ever, I dismissed the worries that were chasing themselves around in my head. The fact that I could do so was the first indication that some of my fears might be rooted in my imagination, not in reality.

Before long, Dave appeared in the cockpit, carrying the Racor fuel filter wrapped in a cloth. As he uncovered it, he explained, "The glass bowl under this filter was filled with black gunk. I figure the rolling of the boat dislodged the stuff from the bottoms of the fuel tanks. We don't have a spare filter, so I'm going to try to clean this one."

With that, he took a deep breath, raised the filter to his lips, and blew long and hard into one end. A greyish-brown froth slowly oozed out all over, soaking the cloth and his beard with diesel. Choking and coughing, he blew into it again and again, until nothing more seeped out. He then replaced the filter, pumped fuel into the lines, and cranked the engine two or three times. It ran! We were underway within about thirty minutes, although it took much longer for Dave to rid himself of the smell and taste of the diesel.

At 1900, after a long day and some 60 nm, we tied up at Hot Springs Cove. Dave dug out the barbeque to cook his salmon, and we dined out in the cockpit, enjoying the food, our surroundings, and the day's accomplishments. In the fading light of the long summer evening, we then followed a two-km-long boardwalk through the rain forest. Heavy cedar planks had replaced about half of the original split-cedar boards, and we noted with interest the boat names and dates carved into the older boards.

With bird songs drifting through the still air, we strolled through woods that were lush and green, and then saw wisps of steam rising above the undergrowth. Finding a stream of hot water bubbling out of a crack in the rock, we followed it down the hillside, seeing it grow bigger before plunging over an eight-foot wall. Below, in a rocky fissure, the stream widened into several bathtub-sized pools before quietly entering the ocean, marking the spot with a steam plume. I couldn't resist the temptation to ease my feet into the hot water in one of the pools, but my enjoyment was short-lived as darkness was falling, and we had a long walk back.

Despite low cloud and fog, we were underway by 0600 on Day 37. The fog slowly dissipated as we motored down the coast, but the day remained cold and grey. With light winds and high swells, the boat rolled heavily for several hours, so we hand-steered. The sea started to flatten out about midday; an hour later, the sun broke through for the first time. Fish were jumping as we passed by the Broken Islands in Barclay Sound, so Dave strung out a line and caught another Coho, a twin to the one he'd landed the day before.

I reached down with a net to scoop up the fish, ducking under the lifeline at the stern as I did so, and caught the brim of my hat. Despite cords securing it back and front, it fell into the water, and I unhappily watched it float away. It was a denim-covered cowboy hat that I rather liked and, seeing my consternation, Dave tried to reassure me, saying cheerfully, "Don't worry; we'll get it back!"

While he took care of his fish, I turned *Windy Lady* slowly onto a reciprocal heading. Without much hope, I started scanning the swells, which were then about twelve feet high with chop on top. To my surprise, I spotted the hat a few minutes later, the small blue crown barely visible as it floated down the back slope of a large swell. I lost sight of it several times in the ensuing minutes, and it then hid behind the bow as we drew closer. Somehow,

I brought the boat just close enough that Dave, stretching to his utmost limits, was able to snag it with the boat hook. As it wobbling precariously on the end of the hook, he brought it back onboard. Although only a small victory, it was definitely satisfying.

We anchored that night on the edge of Barclay Sound, having covered another 60 nm. Day 38 started out grey and cold, bringing fog, rain, occasional high swells, and enough wind to sail for half an hour. About midday, Dave tried fishing again, and soon I heard him reeling in his line. He kept watching and reeling, and I heard him indignantly mutter, "There's a bird after my fish!" Only when the line started to peel out did he realize that he'd hooked a murre (a seabird), and then the line broke.

We were now well into the Strait of Juan de Fuca, and only a day away from the southern tip of Vancouver Island. I kept a close eye on the waters ahead and suddenly glimpsed what appeared to be a breaking wave. Moments later, I saw a waterspout, a tall, black dorsal fin, and another waterspout. Soon maybe a dozen orcas were in front of us, swimming in a long line stretching over to the north side of the strait. As if in a dream, the tall fins broke the water's surface, floated through the air and then disappeared. Captivated, we watched until there was nothing more to see. Another large whale then surfaced three times as it swam by about 100 feet away, providing a marvelous view of tall dorsal fin, shiny back, grey saddle, and black tail.

I continued scanning the waves ahead, and then glimpsed another splash as a whale breached. Before long, many tall, black fins and waterspouts were visible across a large area. The whales were clearly visible, arching in an out of the water, sometimes two together. Dave then shouted and pointed to the other side of the boat, where two orcas were swimming maybe 150 feet away. They were also side-by-side, with the smaller whale emerging from the water just as the larger one re-entered.

I was enchanted by the spectacle but disappointed that they disappeared so quickly; we believe they were hunting Coho salmon. Ten minutes later, another large orca swam past; maybe 100 feet away, it was traveling fast and all we saw was the tall fin arcing in and out of the water.

After a 65-nm day, we anchored at Sooke, on the southern tip of Vancouver Island. We were underway before 0700 on Day 39; it was June 21, the longest day of the year. Dozens of small boats lined the shoreline as we motored up the five-mile stretch of strait to Race Rocks, with many boats clustered around every point of land. It seemed that local fishermen also awaited the salmon.

Passing by the city of Victoria, we continued around the end of Vancouver Island and completed our circumnavigation at 1600, when we dropped the hook at Saltspring Island. That night, we visited with Brian and his wife, and in relating our adventures, first realized what we'd accomplished. We'd simply been too busy living each day to put it all together before then. But as our competence and confidence had grown, so had the attraction of the far horizon.

CHAPTER 5

The Dream Falters

After moving onboard *Windy Lady*, we spent much of the winter poring through cruising guides and magazines, pilot charts and sailing directions. As we processed the information, we realized that it would take two cruising seasons to complete our voyage. The first year, we would sail to New Zealand, where we would sit out the cyclone season. We would then have time for a more leisurely cruise through the South Pacific islands on our way to Australia the following year.

We also had to decide which of two sailing routes to take across the Pacific Ocean. The most popular was through Mexico and French Polynesia, but we read numerous stories about cruisers who made it to Mexico and never left. Not wanting to run that risk, we decided on the alternative, via Hawaii and Samoa.

That decision brought a deadline, as the offshore cruising season in the Pacific Northwest is only three months long. We had to leave by the end of August.

However, the cyclone season in the South Pacific starts in November, and we had 6,000 nm to sail before then, so we really needed to leave as soon as possible. Frantically, we set to work on our "to do" lists, which had grown substantially while we were away.

Five weeks later, on Thursday, July 26, we cleared with Canada Customs in Victoria. We then motored east and north for six hours and tied up at a dock in the American port of Friday Harbor in the San Juan Islands. Here we picked up photocopies of 500 charts of the Pacific Islands and the US west coast that Dave had ordered. We never again wanted to be in a position where we didn't have the right charts.

The following morning was cool, with high overcast, light winds and barely a ripple on the surface of the water. We untied at 0630 and motored northwest through the Strait of Juan de Fuca for thirteen and one-half hours, planning to anchor that night at the western end of the strait. As the tide changed direction twice, our ground speed varied considerably but averaged just under six kt.

I took the helm initially, while Dave spent several hours stowing gear in the cabin. He then found a spot to secure the life raft, where it was easy to access but not too much in the way. Over the course of the day, several small pods of dolphins stopped by to check us out. We always enjoyed seeing them, so would stop whatever we were doing to watch. But that afternoon, a dolphin quietly surfaced, unnoticed, about three feet from where Dave sat reading in the cockpit. It then announced its presence by exhaling with a loud, explosive snort that was so startling, he almost dropped his book.

Later that afternoon, as we basked in warm sunshine, a nuclear submarine came steaming up behind us, its sealed black exterior cutting through the blue waters of the strait. I couldn't see even one sailor, which I thought strange. With my own thoughts swirling with the immensity of the adventure ahead, it suddenly seemed extraordinary that these men would go to sea and never see the ocean.

The sun was dropping low in the west when we turned in toward Neah Bay on the American side of the strait. Near the entrance, we spotted a waterspout and then watched the long, dark mottled shapes of two large, grey whales drifting up the shadowy shoreline. Soon after, we had the anchor down and were secure for the night.

With adverse winds at sea, we spent the next day finishing a few more jobs. Dave spent hours tying netting to the lifelines on the foredeck, so the sails wouldn't slip off the deck when we changed them at sea. Meanwhile, I inventoried and stowed the charts we'd picked up. On Sunday, July 30, we had the anchor up before 0800, and I took a last look around the quiet harbor as we motored out the entrance. I then faced forward, feeling both exhilarated and intimidated by the prospect of sailing some 2,500 nm to Hilo, Hawaii.

Minutes later, near the entrance to the strait, we ran into standing waves that just kept growing higher. With narrow, deep troughs in between, *Windy Lady* was soon pitching almost straight up and down in a way that I really didn't like. Thankfully, we were through the worst of it fairly quickly. We rounded Cape Flattery soon after and, with the ocean heaving around us, stopped to raise the mainsail.

Right from the start, light SW winds were no match for the powerful westerly swells sweeping beneath the keel. With *Windy Lady* rolling heavily from side to side, waves dragged at the stern, turning the bow and knocking the wind from the sails. When the sails flogged, she began to toss and twist, and soon the boom at the bottom of the mainsail was lurching back and forth over the cabin roof.

Realizing that he had to secure the boom, Dave dug out the preventer, a piece of equipment we had not yet used. It was simply a long rope with two pulleys; one end clipped onto a pad eye on the bottom of the boom, and the other end attached to the toe rail. When snugged up, it would stop the boom from swinging, but attaching it in these conditions was no easy task.

Initially, he kept a good grip on the handrail as he started down the side deck towards the mast. But when the boom lurched overhead, bringing the pad eye within reach, he couldn't resist the temptation. Stretching upwards, he tried to snag it with one hand, while flailing around with the other to keep his balance. It didn't take much imagination on my part to picture him going overboard. Looking at the turbulent waters around us, I then realized how difficult it would be to rescue him and sent up a silent prayer.

After several unsuccessful attempts, he continued on to the mast where he was better able to hang on. Stepping up onto the cabin roof, he crouched beneath the boom and stretched his free arm toward the pad eye. Just as he was about to clip on, the end of the boom shot up eight feet into the air, and it swung out over the water on the port side. While we gaped in disbelief, the wind then played with the sail and the boom danced beyond our grasp.

I was still gawking when Dave returned to the cockpit and furled the headsail. He had me turn the bow into the swell, which brought the boom back over the deck. As he eased the mainsail halyard, the sail sagged, dropping the end of the boom, and he then secured it to the windscreen. I could now see the traveler lying uselessly on the cabin roof and realized that the pad eye at the end of the boom had sheared off.

The traveler was another system of ropes and pulleys that controlled the boom and the mainsail; without it, our voyage was over. I'm sure we could have jury-rigged something in an emergency, but it didn't make sense to do so now. After Dave started the engine, I slowly brought the boat around and headed back to Cape Flattery. I couldn't believe what had happened and told myself, "Better here and now, than later, out at sea," but took little comfort in the words.

We were out only six hours and when we returned, the water at the entrance to the Strait of Juan de Fuca was flat, like a large lake. As it was then high tide, I assumed the steep waves we'd encountered earlier had resulted from ocean swells sweeping up the channel at low tide. We motored all night, in and out of fog, and cleared with Customs in Victoria the following morning. Anchoring in the harbor, we removed the boom and took it ashore in the dinghy. The days then dragged by, and we waited nearly two weeks for repairs.

After spending another night at Neah Bay, we motored around Cape Flattery again on Tuesday, August 15. With southerly winds of 15 kt, we happily raised the sails, content to have our plans back on track. By noon, gusty 20-kt winds were taking us a little farther offshore than we really wanted to go. An hour later, with the sun poking through holes in the clouds, the winds began to ease.

By 1500, *Windy Lady* was rolling heavily in swells and winds were down to 5 kt. With the preventer now in place, Dave stood at the helm and hand-steered. He had tried using the autopilot, but in these conditions, it groaned long, loud, and often. As it consumed a lot of power, he decided not to use it under sail. To do so would mean running the engine more and increased the risk of dead batteries.

An hour later, we officially started our watch schedule. I stayed in the cockpit, responsible for keeping the boat safe and on course, while Dave went below to rest. We would share this duty equally and expected to change off every three to six hours, depending on conditions. At that point, I was hoping for good weather for at least twenty-four hours while we ran through the schedule for the first time.

We were now twenty nm off the coast of Washington State, and the mountains on the horizon had started to fade. As they grew smaller and hazier, I found myself glancing in their direction more often. The time soon came when I looked and saw nothing. Uneasily, I stood and turned slowly in a full circle, scanning the waters around the boat. That was when I realized what it meant to go to sea. My entire world had just been reduced to the deck of a forty-foot sailing vessel surrounded by an ocean of moving water.

Within a few hours, I faced another hurdle, one that I hadn't even thought about until we'd discussed the watch schedule a few hours earlier—sailing at night. Hoping that going gradually into the darkness would make it easier, I then requested the sunset watch. As it happened, high swells kept me busy at the helm, so I didn't notice the light fading from the sky. Then, even with high overcast, there was enough ambient light at the end of my watch to see waves closing around the tail of a dolphin. Of course, that was after the animal surfaced behind me, and its sudden, loud snort sent my heart leaping into my throat.

Dave ran the engine for a couple of hours during the next watch, and the dolphins kept him company for half that time. When I relieved him at 0200, we were again under sail; the night was pleasant, with a half-moon and a few stars peeking through the clouds. We were passing through the fishing fleet, so isolated lights bobbed here and there in the darkness. An hour later, the wind dropped and the swell began pushing the boat sideways. With the bow constantly spinning around, I turned the wheel from lock to lock until my shoulders ached. The winds picked up again just before daylight, and *Windy*

Lady settled down at almost five kt, giving us a good run for the next six hours.

About mid-morning on Day 2, an area of white water off the southwest caught my eye. Puzzled, I watched closely and soon saw dolphins leaping in the distance. I called down to Dave and, within minutes, hundreds of animals surrounded us. Most swam with only a dorsal fin breaking the surface of the water; others arched through the air or performed back flips; a few jumped straight up, twisted and landed with a heavy smack on their sides. I tried to count the ones that I could see at any given moment but there were just too many. Fascinated, we watched for over thirty minutes, estimating there were upwards of a thousand animals in the group.

A thunderstorm brought gusty winds early that afternoon, but took the winds when it moved on. With the boat rolling in high swells, Dave then struggled to keep sailing. At change of watch at 1600, he gave up and went down to start the engine. Stretched out on the sole with his head hanging over the engine compartment, he started through his pre-start checklist. But in short order, he was running up to the cockpit, where he spent ten minutes with his head hanging over the lifelines. That was the only time he was ever seasick. With just an hour's sleep the night before, he'd first complained about feeling nauseous when he was in the galley preparing breakfast; diesel fumes and the rolling of the boat then finished him off.

The swells eased just before midnight, so Dave started sailing again, and I managed to sleep for a couple of hours before coming on watch at 0200. With the light of a half-moon reflecting off the waves and stars filling the sky, it was a beautiful night to be at sea. By the end of my watch at 0500, however, the boat was again rolling heavily. When I laid down in my berth, I cradled my body with pillows but still couldn't sleep.

Forcing myself to get up for my 0800 watch, I saw heavy, dark clouds to the northwest when I entered the cockpit. The storm hit just minutes later, bringing a cold driving rain and strong, gusty winds that forced me to stay at the helm. By the time the storm passed, my rain gear was soaked through, I was shivering, and my breakfast was cold. The winds then settled down and we made good time for the rest of Day 3. That afternoon, the color of the ocean changed from the blue-black of northern waters to a warmer, deeper blue.

The breeze died again during my sunset watch, leaving *Windy Lady* rolling in a westerly swell. As waves weren't that high and Dave was trying to sleep, I furled in the jib, put a second reef in the main, and let the boat drift. He then reset the sails when the winds strengthened during the following watch. When I returned to the cockpit at 0200, the sky had darkened ominously, so before he went below, we turned on the spreader lights and put a second reef in the main.

Thirty minutes later, winds were up to 20 kt and *Windy Lady* was crashing into swells, sending spray high over the foredeck. The night had turned pitch-black, and a dense black wall now loomed just yards in front of the bow. Uneasily peering ahead, I had to remind myself that there was nothing in front of us but open water. I then forgot about the wall and concentrated on steering, which proved incredibly difficult using only the compass. I hadn't realized how much I depended on visual cues, like swell direction and cloud formation, to stay on course.

The darkness also hid the arrow on the wind vane, so I attempted to feel the breeze on the side of my face, hoping to detect any change. When a wave washed over the starboard rail, I worried that winds were getting stronger and wondered if we had too much sail up. The last hour seemed to go on forever, but the breeze stayed steady and I made twenty-four miles during my watch.

Winds were steady at 17–18 kt for much of Day 4, so we had another good day. When my watch ended at noon, I stayed and visited with Dave for the first time in three days. Up until then, we'd both spent our off time trying to rest. He was looking tired and said that he hadn't been able to sleep; he was also concerned that we were spending so much time hand steering. He summed it up this way, "It's crazy! We have no time to do anything else! It's also exhausting and we sure couldn't keep it up for long in heavy weather." He then added, "I've decided that as soon as we get into port, I'm going to have wind-vane self-steering installed."

The winds dropped again that evening, and almost becalmed, we drifted through numerous light rain showers. Dave took over the watch at 2200, but woke me at 0030, calling from the cockpit. Turning on the dim light over the berth, I crawled from my blankets and made my way up the companionway. I was still half-asleep,

I guess, so didn't register the bright lights behind him. Waving a hand in their direction, he had to ask, "What do you make of that?"

I woke up instantly as I focused on the mass of lights about one-quarter mile away. Alarmed at how close they were, I screamed, "Why did you come so close?"

He yelled back, "It wasn't me; they came and looked at us!"

Even as he spoke, we heard the hum of powerful engines and the lights started to move away. Visibility was then limited to about a half-mile by a shower that was more mist than rain, so we went below and turned on the radar. The unit took two minutes to warm up, and then a very large image appeared on the screen. It was nearly two nm off and moving away rapidly. While we watched, Dave called repeatedly on Channel 16 on VHF radio, which all ships monitor, but there was no response.

When the vessel was six nm away, we turned off the radar and returned to the cockpit, staring out into the darkness where the ship had disappeared. Dave explained that he'd first seen lights on the horizon about midnight. When he'd looked a second time, they had been much closer, and in spite of his best efforts, the vessel had continued to close. He was exhausted, wet, and very wired—and convinced that it was a UFO. I was just as convinced that there had to be a rational explanation. After all, it was the middle of the night, in the middle of the ocean, and I was coming on watch.

We reviewed what we knew, which wasn't much. *Windy Lady* had been drifting with the waves for hours, and our running lights would have been hard to see in the poor visibility. On the other hand, our visitor had many lights and definitely did not have a diesel engine. The vessel had unquestionably changed course when I turned on the cabin light, and it then sped away into the darkness. As we tried to make sense of what we'd seen, I made a note of the time and place: 0100 on Saturday, August 19, 1995, latitude 43°51.4' N, longitude 126°59.2' W.

Could it have been a US naval ship? Did one pick us up on radar and come over to make a visual sighting? But wouldn't any legitimate vessel have responded to the VHF? Perhaps it was the nuclear submarine we'd seen two weeks earlier in the Strait of Juan de Fuca. If it was running on the surface, maybe it didn't want to be identified; but would it have had all those lights?

The more we talked, the less likely my theories became, so that when Dave went below to try to rest, I stood at the helm, feeling as if I had a bull's eye painted between my shoulder blades. When the mist started to clear, I kept peering around, looking for lights, and heard again in my head the hum of the ship's engines.

To make matters worse, the swells again grew higher, causing *Windy Lady* to roll heavily and knocking what little wind there was from the sails. Once more, I struggled to keep her moving forward, and Dave did the same through the following watch. He hadn't been able to sleep, and with my berth jolting erratically beneath me, neither was I.

Exhausted, I dragged myself up to the cockpit at 0900 on Day 5, telling myself that I only had to make it through the next four hours. Then I saw Dave and my heart sank; he was in worse shape than I was. Knowing that I could continue only as long as he did, I gave my first order, "You've got to get some sleep. I don't want to see you back on deck until you've had three hours. I don't care how long it takes!" He returned two hours later, protesting that he'd just been lying awake in the berth.

By noon, steep ten-foot swells were rolling relentlessly out of the west. With only a light NW breeze, *Windy Lady* rolled heavily, pitching and twisting, and it was an unending battle to keep her under control. Desperately fighting fatigue, we reduced our watches to three hours, then to two, but conditions continued to deteriorate, leaving us pretty much at the mercy of the ocean.

At 1800, we tried jibing the boat, turning to a heading ninety degrees to the east. Dave was then barely able to function, fumbling with the lines as if unsure which one he wanted, and not aware that the preventer had to be moved. We turned back two hours later, as all we'd accomplished was to add a following sea to our other problems.

Conditions reached a new low during my 2000 watch, when winds gusting from 7–16 kt combined with high swells to put ever more stress on the boat. Now, when the sails flogged, the turnbuckles on the stays that supported the mast banged against the chain plates, causing a shudder to run through the entire boat. By the end of my two-hour shift, my shoulders ached miserably from constantly turning the helm, and I realized that I couldn't carry on much longer.

Dave was then a walking zombie, and I knew that I shouldn't leave him on deck, but did. I'd been asleep for maybe forty-five minutes when something woke me. The boat was rolling slowly and abnormally far over on its sides, accompanied by the gurgling sound of fuel/water running over the baffles in the tanks. Disquieted by the unfamiliar sounds and movement, I got up and looked out the companionway. Dave wasn't in the cockpit, and suddenly terrified, I called his name. Out of the darkness behind me, a dull, detached voice mumbled, "The boat's hove-to and I'm monitoring our drift."

He was sprawled on the settee, and I figured that *Windy Lady* had probably spun around on the crest of the waves, backwinding both sails. With no interference from him, she was doing what she was designed to do. I returned to my berth but laid awake until the rolling eased, some two hours later. I heard Dave moving around at 0230, and when he saw that I was awake, he asked, "Can you take the helm?" He added, "The wind has died down, and I want to charge the batteries."

Fifteen minutes later, I was dressed and standing at the wheel, peering into a pitch-black night. It was again difficult to maintain a heading, almost harder with the engine running than it had been under sail. Sometime after 0500, I started to nod off on my feet, so waved my arms and tapped my toes, but was conscious only of a desperate need for sleep. An hour later, I felt the autopilot take control of the helm. Dave then poked his head out into the cockpit and announced, "The autopilot seems to be working and I've had a few hours' sleep. You can come in and go to bed."

I slept fitfully until 0900 and awoke to find Dave standing at the chart table. He came over and sat on the berth beside me, and I saw he'd had a bath and changed his clothes. He now said, "We can't go on like this; we have to get self-steering equipment. I figure we're halfway between Victoria and San Francisco, so we can go to either city to have the work done."

Dismayed at the thought of returning to Victoria, I protested, "If we go back, chances are we'll never leave!"

"Well, that's a risk I'm prepared to take. We know the suppliers and marinas in Victoria, so it makes the most sense to go there. If you want, we could do it in San Francisco, or even hire someone to take the boat to Hawaii."

When I remained silent, he added, "I've already turned the boat around."

Stunned and deeply disappointed, I knew there was nothing more to say. I dug out the charts that we needed to return north and attempted to adjust my mindset. In my heart, however, I knew that the dream was dead. It died at 0900 on Sunday, August 20, after we had sailed 370 nm down the west coast.

Perversely, the sea calmed and the wind hardly registered as we motored north. The autopilot worked well in these conditions, which meant we were both free to relax in the warm sunshine. But that evening, I discovered that the burring noise it made came from beneath my berth. It was like a drill in my ears, so I moved to the settee amidships. The boat movement was somewhat smoother there and I slept soundly for three hours.

As we now had to keep an eye on engine instruments, we spent more time inside the cabin when on watch. We believed in "seeing" rather than "being seen", however, so tried to scan the horizon every fifteen minutes. In that time, a ship making twenty-four kt would travel six nm. As our view of the horizon was only eight nm, that gave us a couple of minutes to take evasive action, if necessary.

The watches that night were very pleasant, with light winds and starry skies intermixed with passing rain showers. Dave finally slept for five hours straight, a good start back to normal. After he relieved me at 0400, I fell into a deep sleep; two hours later, I awoke in a panic, certain I'd been sleeping on watch. Racing up to the cockpit, I stared at him blankly for a few confused moments, and then realized he was still on duty.

Monday was again quiet, with winds of 7–8 knots abaft the beam and a low-moderate westerly swell. When a storm system threatened, we raised the sails but went back to motoring when it passed to the southwest. By noon on Tuesday, an icy wind, steady at 10 kt, was coming out of the north, right on the nose. A US Coast Guard plane found us then, swooping down to read the name on our bow, and called us on VHF radio. They wanted to know our destination, and when Dave explained the situation, asked if we needed assistance. We didn't, but it was comforting to know they were there.

Towards evening, the winds started to strengthen; by nightfall, they were at 15 kt; at 0200, they were at 20.

Wind waves from the northwest were then over three feet and made for a very lumpy ride. Early Wednesday morning, we passed a number of ships standing off river mouths, probably waiting for a higher tide.

About 0900, the breeze backed to the northwest and dropped to 12 kt, so we raised the sails. At 1300, we were thirty-six nm from Cape Flattery, and with winds at 7kt, Dave turned on the autopilot. We started the engine again at 2000 and, two hours later, in total darkness, passed the buoy off the cape.

The trip back through the Strait of Juan de Fuca was long and slow, going against two ebb tides and one nebulous flood. Our speed over the ground hovered near three and one-half kt, but dropped as low as two and one-half. When I relieved Dave at 0115, the night was clear and lights on the freighters easy to see; he had turned on the radar, too, which made tracking the ships much easier. During my watch, I met three outbound freighters, one smaller ship, and crossed a traffic lane.

Just before noon on Thursday, August 24, we tied up at the Customs dock in Victoria. No one showed any particular interest in our early return to port, and a few hours later, we anchored in West Bay. As usual, we waited a while for the boat to settle, but soon were fast asleep.

CHAPTER 6

The Dream Reborn

Next morning, I awoke to the sound of water lapping softly against the hull as *Windy Lady* swayed gently on the anchor chain. Getting up and making myself a cup of coffee, I stood in the cockpit and looked morosely around the quiet harbor. The contrast to past days was neither welcome nor wanted. Although I didn't know what the future would bring, I was bitterly certain that we would never go to sea again. I felt disoriented, partly because the last nine days had been all consuming, and partly because for months our focus had been on preparing to go. In many respects, we'd already cut ourselves off from life in Canada.

Dave now found a berth for the boat at a marina in Victoria's outer harbor, and we began the task of picking up our lives ashore. Two weeks later, we packed up some camping gear and left for Saskatchewan, where we met up with his two brothers in Chamberlain. Honoring their mother's last request, the three men spread her ashes on her father's grave, although strong westerly winds seemed bent on blowing them into Manitoba. I was intrigued by her request, as she'd lived in Victoria for twenty-two years, her husband's ashes were spread in nearby Esquimalt harbor, and her sons all lived in BC. For her, however, home must have been the area where she grew up and raised her children.

That winter, Dave spent time reconnecting with his older brother, Gordon, who lived in Victoria. They'd been very close as youngsters but had drifted apart after leaving home. He also flew east to Ottawa and visited with his son and family for a week, then traveled to Vancouver to be with his daughter when she had surgery on her shoulder. After Christmas, we made a trip north and saw my parents. My mom was not well, although she said nothing. Life, apparently, is what happens despite the plans you make, and at the end of January, we learned she had terminal cancer.

A year earlier, when I told my parents of our plans to sail to the South Pacific, she had shaken her head in disbelief and asked, "What do we do if there's an emergency?" I had responded, "You'll just have to have it without me." Those words now haunted me. I saw her for the last time in April and spoke to her from Hawaii in July and Tonga in September; she died in October. It was a very difficult year for my family members—and for me.

Dave qualified as a HAM operator that winter, while I took correspondence courses on Celestial Navigation and Coastal Cruising. At the same time, he studied everything he could find on reefing systems, boomvangs, and wind-powered self-steering systems. Somewhere along the line, we started making lists again, and he eventually ordered a Hydrovane self-steering system.

We sailed frequently, trying to build on what we'd already learned. Our marina couldn't have been more convenient, as ten minutes after untying we'd be raising the sails out in the Strait of Juan de Fuca. To our surprise, we awoke one morning to find a couple of inches of snow on the deck and pancake ice in the waters around the boat. In mid-April, we set out on a ten-day cruise, staying out in the middle of the Strait of Georgia during the day to get the strongest winds. While the sailing was good, the temperatures were cool, and we turned back about 100 nm up the strait.

Having decided upon the rigging changes he wanted, Dave met with a sail maker in Victoria. The changes were extensive, requiring removal of the mast, and they agreed on a price and timetable. On May 3, we motored to a boatyard about four hours away, where a crane waited to remove the mast. The boat was then propped up on the hard stand, and we camped onboard for the next eleven days; it was not an enjoyable experience. *Windy Lady* quickly resembled a disaster area, both inside and out, as we washed and waxed the hull, antifouled the bottom, sanded and varnished hatches and a hundred other tasks. Our last job before she went back in the water on May 14 was installing a Hydrovane self-steering unit.

The crane was booked for the following morning, but we now learned the mast wasn't ready. That was a major problem because we couldn't proceed with much of our work until it was back in place. Dave badgered the rigger and managed to get it delivered later that day, but we waited around for hours. We then spent three days tied up at the work dock, frantically finishing tasks that had to be done before the boat could be moved back to the marina.

Late in the afternoon on May 18, we untied *Windy Lady* and I took the helm while Dave worked on the GPS. We were about ten minutes away from the dock, when he hollered up, "We've got a problem! Water is pouring onto the galley floor out of the head. I think it's coming from the sink drain."

I immediately wanted to return to the dock, or at least get back into shallow water before we sank, but a closer look revealed the water was coming from a sink tap in the head. One of us had opened it when we were on the hard, and then Dave accidently switched on the water pump while he was working on the GPS. I have to say, nothing got the adrenalin running quicker than the sight of water pouring into the boat.

The next three weeks were a nightmare as we tried to complete the countless and diverse tasks on our lists. What made matters worse was that Dave spent half his time chasing after the contractor, trying to get him to finish his work. We now had time constraints, too, as Brian was going to accompany us to Hawaii. Planning on a departure date near the end of May, Brian had arranged to take the month of June off work.

As we counted down the final days of May, *Windy Lady's* lockers were bulging, and Dave had raised the waterline by six inches. The supplies that we tracked down and stowed away were astonishing. There were power and hand tools; trays of nuts, bolts, screws and washers; miscellaneous engine parts and boat fittings; fuel and oil filters; repair kits for sails and fiberglass; and all the other paraphernalia deemed necessary to maintain the boat, the sails, and the engine. These were all stored in large lockers under the settee in the main cabin.

We had also purchased bags of rice, macaroni, breakfast cereal, flour, powdered milk, baking supplies, and everything else needed to run a kitchen. I packed all the food supplies into large plastic buckets with tight lids to protect them from humidity and water leaks. As I could only keep about a month's worth of food supplies handy for the cook, the bulk of these goods were hidden away in large lockers in the galley. We then made the time to home-can three dozen jars of meat and vegetables and obtained a couple of cases of tinned salmon and canned tomatoes. We would not be using the refrigerator at sea because of the power it consumed, so had fresh vegetables for only a few days.

We had three extra sails, which took up a lot of space, but had their own locker near the bow. Spare halyards, sheets, and anchor rode were stuffed in beneath the companionway steps. After inventorying the charts, I had rolled them up and stored them in sections of PVC pipe, which were stowed beneath the chart table. We had a wet locker for coats and boots, while items like engine oil and filters were stored in smaller lockers in the sole. There was some storage room in the head and in each of the stern cabins, for clothing and personal items. Then there were things like pocket books, equipment manuals, my portable sewing machine, material for tarps, first-aid supplies, and so on, all of which needed to find a home.

Dave finally gave up on the contractor and installed the new deck hardware himself, but the pressure continued to intensify as new sails didn't fit, and the self-leveling radar mast was lost while being shipped. When it did arrive and was installed, it interfered with one of the running backstays needed for the new inner forestay. (*Windy Lady* was now cutter-rigged, so we could raise a third sail.)

On June 4, Dave signed off on everything except for one headsail. The next day, the two brothers moved the boat to the public dock in front of the Empress Hotel in Victoria's inner harbor. On the evening of June 6, we took delivery of the last sail. We were then three weeks behind schedule, and Brian had used a week of his holidays.

Giving ourselves a day to sort things out, we set our departure for Saturday, June 8; if we didn't leave then, Brian wouldn't be coming with us. With no time for sea trials, Dave figured on checking out the new equipment during the sixty-nm run out the Strait of Juan de Fuca. If any problems arose at sea, he was confident we'd be able to handle them now that we had a third crewmember.

We checked the marine weather forecast at 0600 on Saturday and heaved sighs of relief; the day promised to be fair with favorable winds. In a final flurry of activity, we set about readying the boat. Gordon and his wife arrived at 0830, and I was then trying to stow meals I'd prepared for the next four days. An hour later, I heard the sounds of excited voices and laughter, as Brian appeared with his wife and daughter.

At 1030, Dave and Brian loaded the dinghy onto the foredeck, and we were ready to cast off the mooring lines. We all climbed down onto the dock for some last minute pictures, and someone asked me if I was excited. After so many weeks of intense pressure, I just felt numb and glumly answered, "I've been so busy that we'll probably be halfway to Hawaii before I notice that we've untied."

Gordon, who'd spent over thirty years in the navy, asked us to gather round, saying he'd like to bless the boat. He read a passage from the bible, followed by the Twenty-Third Psalm, and then the blessing. Presenting the bible to Dave, he explained that their parents had given it to Granny Ball for Christmas in 1937. I was touched by the brief ceremony, and it remains one of the few things that I actually remember from those tumultuous weeks.

CHAPTER 7

Passage to Hawaii

Under a grey, overcast sky, Brian and I stood on deck waving goodbye, while Dave steered *Windy Lady* away from the public dock. Crossing the harbor to the fuel dock, we topped up the diesel tanks and then untied for the last time at 1115. As we motored out into the Strait of Juan de Fuca, we heaved a collective sigh of relief; it was Saturday, June 8, and we were finally underway.

The men spent the afternoon in the cockpit, sailing for over six hours as we headed down the strait. But I disappeared below and, with inventory list at hand, opened up all the lockers and began shifting and repacking. The problem was that I had desperately been stowing stuff wherever it would fit for days, and now had no idea where I'd put anything. I also wanted to check that heavier objects wouldn't shift and start banging in the first rough seas. We were well down the strait by the time I was satisfied.

The sun was low in the sky when my first watch started

at 2000, and both wind and sea were calm. I looked rather wistfully at the entrance to Neah Bay when we motored past; somehow, it didn't seem right to be heading out to sea at the end of the day. Although Brian took the watch at midnight, I kept him company until we rounded the marker buoy off Cape Flattery at 0115. With smooth seas and a gentle swell rolling beneath *Windy Lady*, we then headed out into the blackness that was the Pacific Ocean.

With 2,300 nm of open water in front of us, we expected to be at sea for at least three weeks. Actually, the distance we traveled would be farther than that because we couldn't sail directly to Hawaii. If we tried, chances were we'd end up with no wind, in the center of the North Pacific High Pressure Area. Instead, we planned to sail southwest from Cape Flattery until we were 100 nm off the coast. We would then head south to Mendocino Ridge, off Eureka, California, where we would again turn southwest to Hawaii.

Our watch schedule was simple, with each shift being four hours long, and no one ever showed up late. Brian came on duty at noon and midnight, followed by Dave and then myself. Actually, Brian always appeared fifteen minutes beforehand, while I turned up ten minutes before the start of my 0800 watch, as I was curious to see what the new day would bring.

Winds stayed light that first night, so Dave continued to motor throughout his 0400 watch. I managed a few hours' sleep but found the engine noise uncomfortably loud. When I entered the cockpit next morning, I stopped and turned slowly in a full circle, scanning the horizon.

That practice would become routine in the days ahead, as I looked for ships and studied sea and sky for signs of what the weather might bring. That morning, I first noticed that the coastal mountains were no longer in sight and then registered the grey, overcast sky and rolling grey waters.

With southerly winds steady at 7 kt, we now set the sails and Dave shut down the engine on his way to his berth. After tweaking the sails a few times, I settled into a rear seat, enjoying the silence as *Windy Lady* glided forward. My body swayed gently with the boat movement as I studied sails, clouds and waves, and lulled by the soft swish of water against the hull, I gradually began to relax. Two hours later, the winds had increased to 12 kt, and with calm seas, *Windy Lady* flew across the waves. For the first time in nine months, I felt in tune with my surroundings, and my spirit soared. Feeling the magic, the dream was reborn.

At the end of my watch, I filled in the journey log, recording our GPS position (latitude and longitude), as well as course, distance covered, barometric pressure, and wind and sea conditions. We recorded this data at the end of every watch, so that in an emergency, our last recorded position would never be more than four hours old. As navigator, I also plotted our noon positions on a chart, which we kept beside the HF radio. In the twenty-five hours since leaving port, we'd traveled 108 nm (Day 1).

The forecast was now calling for cold, stormy conditions overnight, and a few hours later, the winds picked up and heavy clouds moved in. When I came on watch at 2000, lumpy grey seas were tossing *Windy Lady* about roughly, winds were steady at 16 kt, and dark clouds hung low in the sky. I wore a safety harness over top of my bulky cruising suit and clipped onto a jack line as soon as I stepped outside.

Uncertain of what the coming hours would bring, I settled into a rear seat of the open cockpit and watched the sky grow darker. With wind and sea roaring in my ears, the night turned pitch-black, so that a solid black wall again loomed just yards in front of the bow. I focused on keeping wind in the sails and the bow on course and gradually my apprehension eased. The winds strengthened to 20 kt before midnight, but backed to the SSE soon after and dropped to 12 kt.

By morning, winds were again out of the south and strengthening. At 0800, winds were up to 16 kt; an hour later, they were at 18. As the sea grew lumpier, the boat speed picked up, and suddenly I remembered that we hadn't yet checked out the new reefing system that had been part of the refit. As Dave was then sitting in the cockpit, I suggested that maybe it was time to do so. He didn't even bother to respond.

Two hours later, winds were over 20 kt and *Windy Lady* was crashing into waves as she raced across the sea. With spray flying everywhere and the deck bucking beneath his feet, Dave now made his way forward to the mast to put in the reef. For the next thirty minutes, he hung on grimly with one hand, while manipulating lines with the other. Waves repeatedly broke alongside, dousing him with saltwater, and before he finished winds peaked at 29 kt.

For the next sixteen hours, winds gusted from 20–30 kt, and with only one reef in the mainsail, *Windy Lady* charged across the hills and valleys created by the swells. Like an airplane, she moved in three dimensions; the bow could go up or down, or turn right or left, while the deck rolled from side to side. In rough seas, she did all three at the same time and, as she rolled and twisted, waves crashed over the bow, sending spray flying across the deck. Water was soon leaking in through the head hatch, the galley hatch, and the front and side cabin windows.

The rough, erratic, and occasionally violent movements made it impossible to sleep and difficult to do chores. I soon had a collection of bruises on my legs and hips, while Dave flew backwards across the galley on a couple of occasions. I was thankful that he'd had the foresight to have a bar welded across the front of the stove, because even though it was gimbaled and supposed to stay level in rough seas, the bar was all that kept the pots from sliding off.

Ultimately, a second reef had to be put in the mainsail, so at change of watch at 0400, Dave and Brian were out on the foredeck. With the boat hove-to, they worked under the spreader lights, with the wind howling through the rigging and waves crashing all around.

The winds began to ease a few hours later, and by midmorning, seas were somewhat calmer. I now noticed that the color of the water had changed from a cold blue-black to the warmer, deeper blue I remembered from the previous year. A troop of about fifty dolphins then appeared and raced back and forth alongside the boat for nearly an hour. They appeared to be traveling in pairs and were clearly visible, but seldom leapt out of the water.

By noon, winds had dropped below 10 kt, and when I plotted our position, we were 100 nm off the coast of Oregon, just south of the mouth of the Columbia River. We recorded 94 nm on Day 2, and 106 nm on Day 3.

Brian was feeling more like himself when he came on watch that day. He had informed Dave that he would be seasick for three or four days, and he was; he hadn't eaten a thing and left his berth only to stand watch. Now the brothers spent the afternoon together in the cockpit, brainstorming about sails, seas and boat performance, a routine they followed daily thereafter.

When the winds backed to the north, they rigged the boat for downwind sailing, dropping the mainsail and easing the headsail, which then billowed out in front over the starboard bow. With winds steady at 10 kt over the stern, *Windy Lady* was soon making four kt, but rocked uncomfortably as a westerly swell rolled beneath the keel.

For the first time, the temperature warmed up enough that we removed our cruising suits before eating supper in the cockpit. It cooled off quickly afterwards, however, and I was happy to pull mine back on for my evening watch. The evening was pleasant though, and I studied sea and sky as I sat, swaying with the boat movement, and listened to the soft wash of water against the hull. Then, as the light faded and the waters grew dark, I noticed a small, solitary bird flitting between the waves and realized just how much I enjoyed this last hour of the day.

After three days at sea, we were settling into a routine; standing two watches each day, adjusting sleeping habits, and adapting to living with a limited supply of fresh water. Dave also looked after the engine, kept in touch with a HAM network, and prepared our meals. As well, he monitored the batteries and usually ran the engine for four to five hours every fourth day in order to charge them. I took notes of the weather broadcasts, plotted our noon positions, and kept the galley clean and the cabin tidy.

That night the winds backed to the west and the weather deteriorated. When I came on watch at 0800, it was rainy and cold, and heavy clouds darkened the sky. The early forecast included a gale warning for the southern Oregon coast, so I was worried. A later forecast extended the warning to northern and offshore waters, and I worried even more. Fully expecting 40-kt winds within twelve hours, I went through the galley and tied everything down.

The forecasted winds never materialized, but they did gust up to 25–30 kt for the next three days. They also moved constantly, first slowly veering ninety degrees from west to north, then backing seventy degrees to WNW, and finally veering forty-five degrees to NNW. As a result, sea conditions were extremely rough, and *Windy Lady* raced through the waves with her hull heaving and twisting. Sea state was just as important a factor as wind speed when it came to setting the sails, and we ended up putting in or taking out reefs at every change of watch.

While sea conditions were hard on the boat, they were also hard on the crew, making our watches difficult and sleep almost impossible. But reducing sail was not done lightly as that would slow the boat, and we had a schedule to keep. As a result, the men again worked under the spreader lights at 0400, using the engine to control the boat, while putting put three reefs in the main and partially furling the headsail.

After six full days at sea, we'd sailed 686 nm, having recorded 107, 140 and 135 nm for Days 4, 5, and 6. We were then nearing the western extremity of Mendocino Ridge, some 270 nm off Eureka, California. The winds had taken us farther offshore than expected, and we were concerned that we might be a little close to the center of the high. We didn't know its exact location because I was no longer picking up weather broadcasts. Still, we turned the bow southwest towards Hawaii.

The following morning, the blustery conditions began to ease. By noon, the NW breeze was steady at 15 kt; by late afternoon, it was down to 10 kt. As we went into our second week at sea, we hung up our cruising suits and safety harnesses. Winds varied only a few knots over the next twenty-four hours, while barometric pressure went up and down a point. We logged 109 and 107 nm, on Days 7 and 8.

Our luck then ran out. The barometer climbed five points overnight and the wind died. Come morning, *Windy Lady* rolled in the swells, and every wave that swept beneath the keel caused the sails to flog and the rigging to bang. Before long, a few items had worked loose in the lockers and were hammering back and forth monotonously. It was all I could do to stay at the helm until noon before tracking them down. We recorded 92 nm on Day 9.

Concerned about getting Brian to Hawaii, Dave now spent an hour on the HAM network. After confirming that we really were in the center of the high, he started the engine. As the sea calmed, I took advantage of the quiet, sunny conditions and toured the foredeck, checking fittings and halyards. I wasn't concerned to find a few fittings that needed tightening, but was shocked to discover that the mainsail halyard had chafed halfway through. When Dave later repaired it, he cut three feet off the shackle end. From then on, checking fittings and halyards became part of my regular routine.

As I stood at the bow that afternoon, admiring the sparkling blue waters around us, I had no idea that the ocean floor was almost three nm down, or that the closest land was 700 nm away. I only noticed the depth, 5,225 meters, when I did my chart work next day. However, after nine days at sea, I was beginning to appreciate what an insignificant speck we were on the surface of the ocean.

Then, during my evening watch, I spotted a plastic bottle bobbing on the surface and stared with dismay as other bottles, fenders, and even the upturned bottom of a bucket came into view. Without wind waves, they were easy to spot, and I had to wonder how much other stuff was floating around us.

Another full day then dragged by with the engine throbbing noisily and the boat rolling relentlessly. Still, we made good time under power, and our watches were definitely more relaxed with the autopilot doing the steering. We raised the sails again at 0800 the following morning, after motoring for forty-two hours; winds were then at 5–10 kt from the west. On Days 10 and 11, we recorded 120 and 122 nm.

About then, somebody asked, "Shouldn't we be getting close to the trade winds?" Instantly, we all shared a vision of a steady 15-kt breeze filling the sails. We began to watch the wind vane closely and perked up when the arrow drifted to the northwest. We grew excited when it moved on to the north a few hours later, and actually cheered when it settled in the east. However, winds remained light and variable for two long days. Under sail, we registered 95 and 88 nm for Days 12 and 13.

We had then sailed 1,327 nm and were roughly halfway to Hawaii. With only ten days left in Brian's holidays, Dave again started the engine. An hour later, the air stirred slightly, and by 1400, we were again under sail with the breeze steady at 10 kt from 030 degrees (NNE). We had reached the trade winds.

With winds coming over the stern, the men dropped the mainsail and eased the headsail out over the starboard bow. Later, for the first time, they attached the whisker pole to the clew and stretched the sail out like a big vertical kite. I viewed the configuration suspiciously, thinking it very inflexible, and wasn't too happy when Dave decided to run with it overnight. The problem was that the pole really needed two people to handle it, and I disliked not being able to look after things on my night watch. Fortunately, the winds remained steady and I didn't have to roust anyone out.

We logged 110 nm on Day 14, but as we started our third week at sea, trade winds proved unreliable. Soon after Dave and Brian settled into the cockpit that afternoon, the breeze veered eighty degrees to ESE. With the jib already poled out to starboard, they raised the mainsail and eased it out to port, in a configuration known as wing-on-wing. Later, when winds backed to the east and were abaft the beam, they stowed the whisker pole and reset the sails for a broad reach.

When I relieved Dave at 2000, he reported that he hadn't been able to balance the sails, so had hand-steered for most of his watch. I wasn't too concerned, as winds were then only at 10 kt. By the time I decided that the mainsail really needed a reef, both men were in their berths, so I told myself that I could cope until midnight. That was a mistake. While the first hour was a struggle, the following three hours were a nightmare. The problem wasn't just the wind, which was abaft the beam, but also a strong following sea.

Generally, if we had up too much mainsail in gusty conditions, the bow of the boat would keep turning into wind. I would then fight with the helm as I tried to hold the bow on course. Just how difficult that struggle was depended on the angle of the wind to the sail, as well as sea state. A following sea, however, turned it into a whole new ball game.

For the first hour, the battle was unrelenting, as every time I fought the bow back on course, another gust would start it anew. Then, as darkness fell, winds increased to 15 kt and swells grew higher. Now, as I fought with the helm, the boat spun around like a top on the crest of the waves. The mainsail would blanket the headsail, and when the headsail lost the wind, the boat was uncontrollable. At its worst, *Windy Lady* pivoted over 100 degrees, backwinding both sails. I furled in the headsail and managed to get her sailing again by slowly circling her around using the mainsail. To my surprise, I still made eighteen nm on that shift, although I thought I'd been going in circles.

It was difficult to call for help in this kind of situation. It wasn't just that I disliked the idea of rousting someone from sleep, but frequently, it was impossible to leave the helm. So, I became much more vigilant when I checked conditions at the start of my night watch. Instead of asking myself, "Do we need a reef?" The question became, "Can we forget about putting in a reef?"

Next morning, winds were down to 10 kt and sea conditions had eased; at noon, we recorded 107 nm for Day 15. That afternoon proved to be the nicest yet, and the men spent their time working with the sails, figuring out all the things that we should have known before leaving home. Even better, with lighter winds, they were able to get Otto to hold the boat on course.

Otto was our Hydrovane self-steering system. We'd all spent countless hours adjusting sails and tweaking vane settings, but could never get it to work for more than a few minutes at a time. After mulling over the problem in their afternoon sessions, the men concluded that there had to be slippage in the boat's hydraulic steering system, which was allowing movement in the main rudder.

A few squalls brought gusty winds and rain showers overnight, and next morning, the world had been scrubbed clean. Bright sunshine danced on whitecaps that dotted deep-blue waters, and the day absolutely sparkled. A brisk wind had also arranged rows of small, fluffy, white clouds across the sky. As I did my usual scan of the horizon that morning, I again marveled at the sheer size of the ocean we were crossing, but then immediately thought of the insignificant speck we made on its surface.

Just before sunset, I glimpsed a school of flying fish, their gossamer wings looking like a net thrown out over the waves. During the next few days, the fish appeared more frequently, emerging from the waves and gliding through the air, almost appearing to fly. Subsequently, we found remains on the foredeck from a few that came aboard during the night.

With NE winds ranging between 5–12 kt, we made 105 and 106 nm on Days 16 and 17. As we basked in bright sunshine and warm temperatures, Dave and Brian tried their hands at fishing. When they caught a twenty-pound tuna, they weren't able to lift it into the cockpit, but the four-pound mahi-mahi that came next made a welcome change to the dinner menu. Winds then picked up to 10–15 kt from 030 degrees, and we made 124 and 121 nm on Days 18 and 19.

Except for one day of squally weather with gusts over 20 kt, winds were up and down from 5–15 kt for the remainder of the voyage. We averaged 119 nm/day for Days 20 through 24. We crossed the Tropic of Cancer when we were three days out, and the following morning, I picked up a weather report for the first time in two weeks.

Two large freighters passed two nm behind us early the next day, and by midmorning, we were searching the sky for signs of Mauna Kea, the 13,800-foot volcano that towers over the island of Hawaii. At 1410, we finally made out the mountain's dim shape through the surrounding cloud cover; we were then only thirty nm away.

With winds below 10 kt, we wouldn't make it into harbor before dark, so we spent half of our twenty-fifth day at sea hove-to about twenty miles off. I stood my usual watch that evening and was relieved that the passage was over; it seemed we'd been at sea a long time. As *Windy Lady* drifted slowly in a light breeze, I watched the muted shades of the sunset spread across the sky and thought about the events of the voyage.

The days had long since blurred one into another, but much of the time, we'd gazed out over waters that were usually blue and skies that were often grey. The view had always been the same and yet constantly changed, and I was never bored. On several occasions, we'd seen small pods of dolphins, and most days, I'd seen a bird or two. Occasionally, one had spent ten to twenty minutes trying to land on the masthead, but the VHF antenna on top of the swaying mast had thwarted all efforts.

The last hour of the day had captivated me most, and I'd sat and watched the light fade from the sky as night's shadows crept silently across the water. Often, as the sun sank into the sea, it created a golden pathway to the horizon, while the rising full moon had created a silvery pathway of its own. On quiet nights, the stars had cast their reflections on the water, while one stormy midnight, I'd been astonished to see what can only be described as the shadow of a rainbow.

It occurred to me that such scenes would have been familiar to the early mariners who sailed these waters, to the crews of the Spanish galleons that had crossed from Mexico to Manila some 200 years before James Cook arrived in Hawaii in 1778. While they'd had only their skills as seamen to keep them safe, our passage was made easier with modern equipment and electronic devices. Still, we shared with them the same vulnerability to the whims of wind and sea; if bad weather developed, there was no place to hide. And I had worried about bad weather, sure that in three weeks at sea, we were bound to run into a storm.

I also brought personal baggage with me and, in the lonely vigils of my evening watch, was obsessed with thoughts of home and family. All the things that I had buried during those last frantic weeks had risen to the surface. I knew I would never see my mom again and felt I'd deserted my dad and siblings when they needed me most. As I wrestling with feelings of guilt and loss, I also sensed that my relationships with all of them would change.

I took what solace I could from the natural world around me, but came to resent the presence of a third person on the boat. I knew I was being completely unreasonable, but couldn't help myself. For days, I went around with a black cloud over my head, so kept to myself because the last thing I wanted to do was disrupt the passage.

Late one afternoon, well into the second week, I stood on deck, staring out at sea and sky. I don't know what happened, or why, but suddenly the view in front of me seemed extraordinary. That I could actually be sailing across an ocean was unbelievable, an adventure beyond my wildest imaginings. But even as I marveled at my incredible good fortune, I realized that in my despair, I had likely missed huge chunks of it.

The thought of wasting even a moment of this once-in-a-lifetime adventure was more than I could bare. So, I started another journey, one in which I tried to live each day as it came. Knowing that I could not change what had already happened, I fought to put the past behind me. I also tried to quit worrying about the future, because we were as prepared as we could be for whatever was going to happen.

I wish I could say that I also came to terms with Brian's presence, but that only happened when I acknowledged how much easier the passage was with his help. Much later, I felt the need to explain and apologize. With his usual sunny good-humor, Brian just shrugged it off, saying he hadn't noticed.

That night, as we hove-to under a triple-reefed mainsail, the boat drifted eight nm west and two nm north. Next morning, an enormous black cloud hung over

Hilo harbor. Dave took the helm when we set the sails at 0700, and two hours later, started the engine in order to charge the batteries. I had studied our photocopy of the harbor chart, coloring in the hazards, and steered *Windy Lady* into the anchorage at Radio Bay with no difficulty. Twenty-six days after leaving Victoria, we dropped the anchor at 1130 (0830 local time), having sailed 2,756 nm.

CHAPTER 8

Hilo and the Big Island

We stopped in Hilo when we arrived in Hawaii because it was the closest port of entry to our route. As captain, Dave now had to report to authorities ashore, so we lowered the dinghy over the side as soon as we finished anchoring. Pulling on long pants and a shirt, he gathered up boat papers and crew passports, and rowed ashore. When he returned two hours later, he had dealt with Immigration, Customs, Agriculture, and Port Authority.

Meanwhile, Brian and I remained on the boat, sweltering in the unfamiliar heat and humidity. He was anxious to book his passage home, as it was July 3 and his holidays were over, so spent the time readying his belongings. I kept busy tidying the deck, even scrubbed the salt off during a torrential downpour. As soon as Dave returned, Brian dinghied ashore and booked an airplane seat for next day. After seeing him off to Canada,

Dave and I had ringside seats for the harbor fireworks celebrating the American Independence holiday.

We intended to stay in Hilo just long enough to rest up, but quickly learned that living on a boat, in summer, was not nearly as enjoyable as a winter vacation in a nice hotel room in Waikiki. The temperature inside the cabin was ninety degrees plus, and the oppressive heat and humidity left me feeling bone weary and unable to sleep. Frequent, heavy rain meant that we couldn't leave the deck hatches open, and without shore power, we couldn't even plug in a fan. Because of the rain, we were also bailing out the dinghy three or four times a day.

After four days of misery, a position opened up on the retaining wall, where there was room for maybe twelve boats to stern-tie. Moving over to the wall meant that we could rig tarps over the two forward hatches and leave them open, so we wasted little time, dropping the anchor at the edge of the channel and backing into the berth. We then tied stern lines to the wall, which was far enough away that we had to ferry across in the dinghy. The open hatches increased airflow inside the cabin and before long had reduced the temperature by ten degrees.

For a moorage fee of $8 US/day, plus a $50 deposit for the key, we had access to a small ablution block containing toilets and showers. Within a few days, we'd also located a coin-operated laundry nearby. We ate our main meals ashore, as dumping almost anything in the harbor could bring fines. Within a week, we were walking five miles a day, sometimes farther. The rain didn't deter us, as the air cooled off during the frequent downpours, and our clothing would be dry in half an hour. Of course, it wasn't quite so convenient to wait out downpours with a load of clean laundry, or after a shower.

In the late afternoon, crews off cruising boats often gathered at a picnic table set up in a grassy area near the retaining wall. Most of the yachts were crewed by couples going back to the Pacific Northwest from Tahiti, and for the first time, we heard first-hand information about the countries and harbors that we hoped to visit.

At the same time, I began a running battle with sand fleas or some other such pest that hid in the grass, and small, red bites soon covered my legs and itched for days. Fruit flies/black flies were a problem in the galley, too, and a few crumbs on the floor would start an infestation. We started keeping our fresh fruit out in the cockpit, but I still occasionally grabbed a fly swatter and cleaned house.

The areas where we walked didn't seem very prosperous, although the streets were busy and vehicles constantly whizzed past. There were only a few people in the malls, and stores were mostly empty, so clerks tended to pounce on customers. Although I searched for the sights, sounds, and smells I remembered from earlier visits to Hawaii, they seemed buried beneath the roar of commercial activity.

We learned that the sugar cane industry had shut down because it could no longer compete on world markets. A local teacher told us that the majority of her students came from homes on welfare, with the associated problems of alcohol use and wife and child abuse. Settlements for land claims by native Hawaiians had also stalled, and they had taken over much of the beach area around Hilo. A ramshackle settlement had grown up next door to Radio Bay and, on weekends, we heard loud voices as people gathered and partied.

In fact, the anchorage at Radio Bay was tucked into a small corner of Hilo's busy harbor, with a cement plant on one side and a container dock on the other. With trucks and equipment working nonstop, and the occasional aircraft taking off, it was a very noisy place. We talked about sailing to Maui or Kona, but decided that didn't make sense, as we had to check out of Hilo when we left.

One afternoon, we returned to find the marina in an uproar, as a squall with 30-kt winds had caused havoc. One of the boats, *Compass Rose*, pulled up its anchor and drifted into *Swak*, which then ran into *Irish Lady*. *Compass Rose* attempted to motor away but ran afoul of *Swak's* mooring line, which then wrapped around its prop. The crews were still trying to sort them out when we returned, but our neighbor told us that *Windy Lady* barely moved, which made us happy.

Dave rented a car during our second week, and we drove up to Kilauea Volcano, which had then been erupting for some twelve years. After peering into Halemaumau Crater, we drove the Circle Rim tour, but visibility was poor, with steady rain combining with steam from the vents. Conditions improved as we drove down the Chain of Craters Road, which emerged high up on a range of hills, providing a spectacular view of a broad, flat plane beside the ocean. The view was of even more interest because of recent lava flows.

The flow of molten rock followed the contour of the hills, dropping down a couple of hundred feet as it did so. It then crossed the flat, spreading out over large areas of countryside and sending up a huge steam plume as it fell into the sea. The plume was visible for miles but, as we drew closer, dissipated into smaller columns that identified many sites where lava streams entered the water. Parking the car, we walked over new lava to the ocean's edge, marveling at this display of nature's power. In all honesty, I also recalled the spooky tales of Madam Pele, the volcano goddess, which I'd heard from Hawaiian friends in times past.

The current lava flows originated from a cone on the mountainside and reportedly flowed through a lava tube for six miles, so we visited a section of older tube that was then a tourist attraction. Standing out in the rain, I studied the story boards, which explained how the outer layers of lava had cooled, forming a pipe that allowed the inner core to drain out through the center.

Climbing down into the entrance, we walked through the impressive tunnel for about two hundred feet. Lamps placed along one wall provided the only lighting but it appeared to be about fifteen feet high and twelve feet wide. With lamplight glinting off small pools of water on the floor, roots hanging down from above, and shadows all around, I found it just a little spooky.

Our last stop that day was in a small community on Hilo Bay, where a tall clock stood in a park with hands frozen at the time a tsunami hit in 1960. At a nearby café, lines on a wall marked the height of the water that day, as well as the height of a previous tsunami. Impressed as I was by these reminders of past disasters, it never occurred to me that one day we would be on the periphery of just such an event.

Keeping the car a second day, we toured the northern half of the island and spent so many hours sitting that we were both saddle-sore and weary on our return. We drove through abandoned cane fields, crossed through dry grasslands and saw a lot more lava. We also admired views of isolated valleys, sandy beaches, and magnificent ocean scenery.

The following weekend, I found traces of the old Hawaii, when we happened to wander by a downtown park just as the International Festival of the Pacific was getting underway. Drawn by the lyrical sounds of Hawaiian music, we stopped for an hour and watched hula dancers performing on the beach. The first group was from Japan and the dancers were obviously professionals, but several local clubs followed and their members were not quite so polished. The littlest ones had more enthusiasm than grace but proved to be just as entertaining.

Toward the end of our stay, a small fish boat tied up beside us. The crew took the boat out every evening, worked all night under lights, and returned in the morning with their catch. They were after yellow-fin tuna and never brought in more than four fish. But the fish were impressive, with two of them weighting in at over two hundred pounds apiece.

After two weeks in Hilo, we started to think about the next leg of our journey, which would take us 1,000 nm south to the coral atoll of Palmyra. We trekked downtown to a realtor's office and applied for a visitor's permit, giving July 23 as our departure date; we then returned a few days later to pick it up.

Our maintenance work was well underway at that point. Working between rain showers, we'd replaced the sealant around the cabin windows, which was a big job. Dave had also found a leak in the hydraulic pump at the inside steering station. Having disconnected it, he was hopeful that he'd solved Otto's problem. He next discovered that the mushroom vent on the head hatch leaked because it wouldn't fully close. He fixed it by removing the bolts and inserting them from the opposite side.

The furling drum on the head stay was now frozen, and we poured buckets of water through it. When that didn't help, Dave sprayed it with WD40 and, eventually, metal filings washed out. Not a good sign! I climbed the mast a couple of times, attaching a halyard to my climbing belt so Dave could tie me off. I first replaced the cups that were part of the wind gauge, as one had broken off after a week at sea. I then re-attached the radar reflector to the top of the backstay, as it slid down during the crossing.

We re-filled one propane tank and replaced supplies, then bought material to make flags. When we arrived at a port of entry, we were supposed to fly a Canadian flag, the flag of the country we were entering, and a yellow quarantine flag (Q-flag). I put that project on hold, however, when I discovered that a dozen jars of home-canned meat were covered in mold. Disgusted, I wiped them off and washed them down with hot water and bleach. I figured the humidity had attacked a film of grease left on the jars when I removed them from the canner. As the jars were sealed, the contents were fine.

On Monday, July 22, the weather outlook wasn't promising; high swells had reached Hawaii from a storm off Tahiti the previous week. The forecast was calling for several days of SE winds, and all south-facing shores were under high surf warning. While mulling over whether or not to leave the next day as planned, we went up on deck and found the dinghy missing.

Anxiously looking around, we could see no sign of it anywhere. About the time we started to wonder what to do, a sailboat anchored behind us swung with the tide, revealing two dinghies tied alongside. Keeping our fingers crossed, we waited. Bob, a single-hander on *Adios*, brought it back around 0900, cheerfully saying, "I found your boat drifting out to sea last night."

Deciding not to fight headwinds from the start, we delayed our departure, but then the time dragged and I started to have second thoughts. Brian wouldn't be with us on this leg, and we would be heading into a very isolated region in the central Pacific. I began to wonder just how much help I would be in an emergency. Needing to distract myself, I kept busy scrubbing deck and sole, then baked a large batch of ginger cookies, as I'd heard that ginger helped settle upset stomachs at sea.

Dave kept busy, too, and made a couple of half-hour round trips, jerry-jugging diesel fuel from a nearby gas station. This was an anxiety-filled chore as fuel spills, even small ones, were subject to $10,000 fines. He also installed two small, 12-volt fans, one in each berth. On Wednesday afternoon, we checked with Customs and learned that the officers would be in Kona the following day, so obtained our exit clearance.

Returning to *Windy Lady*, we prepared her for sea, taking down canopies, filling water tanks, and securing life raft and buckets. That night, for the first time since we arrived, the temperature inside the boat fell below eighty degrees. When the early forecast next morning called for easterly winds of 10 kt, we raced through the boat, checking that everything was secure inside and out. We returned the washroom key to the Harbor Master's office, recovering our $50 deposit, and untied the mooring lines.

CHAPTER 9

Passage to Palmyra

We left Radio Bay at 1040 on Thursday, July 25, with Dave winching the dinghy aboard and stowing away the fenders while I piloted the boat through the harbor. Twenty minutes later, *Windy Lady* was rolling heavily as large swells hit her broadside when we passed through the breakwater. She settled down quickly once we turned on route, so we stopped and raised the mainsail. We then continued under power, charging the batteries, while we followed the shoreline around to Hawaii's East Point.

The crossing to Palmyra was really the first leg of a 2,400 nm voyage to American Samoa. The atoll was about 1,000 nm south of Hawaii, and we expected to be at sea ten days. No one lived there, other than the caretaker, although a few yachts did occasionally visit. We now knew that the US Navy dredged the channel into the lagoon during WW II. We had GPS coordinates for the entrance, which reportedly was marked by a buoy with a range marker on shore.

As temperatures were over ninety degrees Fahrenheit, we spent the afternoon in the rear seats of the cockpit, where a light breeze somewhat tempered the heat and humidity. Bright sunlight glinted off the green slopes and dark waters around us, but clouds covered the summits of both Mauna Kea and Mauna Loa. Steam venting from Kilauea's volcanic cone added an exotic touch, drifting downslope and eventually mixing with plumes rising from the lava that was building new land south of the point.

Before long, however, Dave was fretting about the batteries, which became overly hot and took longer than normal to charge. An energy monitor (EMON) controlled the high-output alternator, and he worried that the technician in Canada had set the charging rates too high for the tropics. If so, our expensive, new gel-cell batteries could be damaged.

We finally shut down the engine at 1600 and started sailing with ENE winds of 10 kt and a two-foot swell. Minutes later, we turned the bow southward and began to draw away from the coast of Hawaii. *Windy Lady* rolled easily through the waves for the next two hours. Lulled by the gentle rocking of the boat and soothing swish of water against the hull, we sat and absorbed remarkable vistas of open sea and sky ahead.

When the light started to fade, Dave thought of his coming night watch and went below to rest. I now turned and looked back, studying the island as it drifted into the horizon. As evening shadows draped the landscape, I found myself straining to keep the details in focus, and when a few pinpoints of light appeared, I couldn't tear my eyes away. Then, when the last one disappeared, a feeling of dread swept through me.

The doubts that had nagged at me for the past few days suddenly swirled around in my head, and I was terrified. We were in a very small boat heading into an isolated area of a very large ocean. There would be no place to seek shelter, no one to call for help. Was I really ready for this? I then looked around at the quiet waters and dark sky, and knew that whatever happened, this was where I wanted to be. With that, I brought my imagination under control and the turmoil subsided.

Dave relieved me at 2000, and we changed the watch every four hours thereafter. We kept to that schedule because neither of us could sleep longer than three hours at a stretch when we were at sea. Still, it took some getting used to, as we were getting up and going to bed three times a day.

When I returned to the cockpit at midnight, the night sky was bright, with the light from a three-quarter moon shimmering on the water. ENE winds were still steady at 10 kt, and *Windy Lady* was sailing smoothly through a ruffled sea. Two hours later, the moon dropped behind the horizon and stars filled the sky, with the brightest reflected on the ocean's surface. The glow of the Milky Way then spread across the heavens from northeast to southwest. As if that wasn't enchantment enough, tiny flashes of light now sparkled in the water as the passing of the boat disturbed some form of bioluminescent organism.

Just before change of watch at 0400, the winds picked

up, bringing a squall with 25-kt gusts and rain showers. The storm wasn't surprising as we were heading into the Intertropical Convergence Zone (ITCZ), an area north of the equator that formed a buffer between northeast and southeast trade winds. The ITCZ, also called the Doldrums, is notorious for its swift tropical storms and dead calms. We would learn that the worst storms hit in the middle of the night, while calms set in during the heat of the day.

When I came on watch at 0800, winds were from the east at 10 kt, but squalls regularly raced down on *Windy Lady* throughout the morning. Although now working better, Otto couldn't handle the 25-kt gusts, so I learned to ease out the main and partially furl the jib before the winds hit. That took most of the pressure off the helm, and it was then a simple matter to reset the sails afterwards.

About midmorning, the winds increased to 15 kt and veered to the SE, putting them on the bow. I was then steering fairly close to the wind and didn't pay enough attention when the next squall hit, so backwinded both sails. With the boat dead in the water, I had to circle around using the mainsail in order to start sailing again. At noon, after twenty-five hours at sea (Day 1), we logged 119 nm.

For the next forty-eight hours, winds shifted back and forth from east to southeast, while fluctuating from 10–20 kt; they frequently moved thirty degrees during a watch. Squalls also brought 25-kt gusts, periods of rain, and even pushed up six-foot seas on occasion. The daylight hours were hot, humid, and not very pleasant, but the nights were pure magic, with a waxing moon lighting up the water, bioluminescence sparkling in the shadows, and cooler temperatures. We logged 114 and 128 nm for Days 2 and 3.

The winds backed to the northeast that morning, and four fast-moving squalls brought gusts of 30 kt. The winds then veered eastward, growing stronger, and by late afternoon, heavy grey clouds filled the sky. By midnight, winds were steady at 20–30 kt from the east, putting *Windy Lady* on a beam reach. Even with three reefs in the mainsail, she raced across the sea, tossing and twisting, and crashing into swells. Before long, the cabin windows were again leaking, the water seeping in more easily than ever, and the wild movements of the boat dispersed it everywhere.

When night reluctantly gave way to dawn, there was no horizon, just a leaden sky merging into a dull, grey sea. Swells, ten-twelve feet high, broke alongside, rocking the boat, with spray flying everywhere. A grimy layer of salt soon covered the deck and cockpit, and then tracked below. At noon, we recorded 133 nm for Day 4; winds were then on the beam at 15–25 kt.

Conditions remained much the same for another twenty-four hours, with bursts of white spray providing the only relief to the grey monotony that surrounded us. Boat movements were unpredictable, sometimes even vicious, and I soon had a new collection of bruises. To add insult to injury, that evening I had to scrape my dinner off the cockpit sole after a moment's inattention.

Inside the cabin, heat and humidity were almost

unbearable, and sweat poured from our bodies at the slightest exertion. Other than preparing meals and keeping the galley clean, we limited our duties to the essential. For Dave that meant spending thirty minutes each night on the HAM network, while I did whatever calculations were necessary in order to plot our noon positions on the chart.

We were spending maybe an hour together each day, during breakfast and supper; otherwise, we tried to rest when off duty. I made a nest in the port berth, using two fenders and several cushions, which softened the boat movement enough that I could sleep. Still, it was only with air from the small 12-volt fan blowing on my face and shoulders that I was able to do so.

We logged 131 nm on Day 5, by which time winds had dropped to 15–20 kt. Seas remained high, however, and grew rougher. As Otto couldn't cope, we now hand-steered. By midnight, ENE winds were down to 10–15 kt but seas remained erratic. Eight hours later, Dave recorded his frustration in the journey log, writing, "The vane, the vane, it's driving me insane!" The morning brought squalls and more rain, and at noon, we recorded 118 nm for Day 6.

The wind died about mid-afternoon, leaving *Windy Lady* rolling uncomfortably in rough seas. Thirty minutes later, it rose out of the SSW, right on the nose at 8–10 kt. We reset the sails but were unable to steer our course, so gradually drifted westward. About midnight, a storm brought winds of 18-22 kt and two hours of steady rain; when it cleared, the winds again died. We fought to keep sailing until 1000 next morning, when Dave gave up and started the engine. Fifteen minutes later, the autopilot quit working and we were back to hand steering. He later discovered that water from the leaking windows caused the malfunction. We logged only 80 nm on Day 7.

A light breeze stirred the air about 1700, and an hour later, we were again under sail. At 2000, NE winds were steady at 8–12 kt and lightning was flashing in the night sky. Four hours later, the night had turned pitch-black and rain was pouring down. All through my midnight watch, winds gusted from 15-25 kt, while squalls brought 30-kt gusts and heavy downpours. Fortunately, Otto only needed the occasional assist. Winds then veered to the east and grew stronger, and Dave faced gusts up to 39 kt and turbulent seas during the first hour of his watch.

At 0800, winds were down to 10–15 kt; two hours later, they were again on the nose and pushing us westward. We recorded 92 nm for Day 8. That afternoon, we turned to an easterly heading for a while, but tacked back when we ended up even farther off course. We now held the bow as close to the wind as possible, willingly trading speed for direction.

Knowing we would have to make up any easting we lost, we kept a close eye on two of our instruments. The Signet measured the distance the boat sailed through the water, while the GPS measured the distance made good towards our destination. For the four hours ending at 1600, we sailed twelve nm through the water, but made good only five nm towards our destination. At 2000, we sailed another ten nm but again made good only five nm.

We grew more and more concerned as the hours passed, but when I saw the numbers at midnight, I was downright spooked. We had sailed eight nm during the previous four hours, and were a mile farther away from our destination. It seemed impossible, like a big, black hole was sucking us backwards, or something was affecting our instruments.

I stared at Dave in shocked disbelief, and he then moved over and started the engine. But then, with the throttle set at 2000 rpm, *Windy Lady* crept forward at only two kt, instead of the usual four to five. As well, we now noticed that the course shown on the GPS varied by thirty degrees from the compass; they were always identical! Standing in that dimly lit cabin, in the middle of the ocean, in the middle of the night, I suddenly felt very lonely.

Five minutes later, the course readings came back into sync and the boat speed picked up a knot. We then realized that we were in the grip of strong currents, much the same it turned out, as if we'd been crossing a west-flowing river, a hundred miles wide.

As it was now my watch, I took the helm while Dave went below to rest. With the engine throbbing in my ears, I hand-steered, peering out into a night that just grew blacker as we headed south. We averaged only two and one-half kt during that watch. Dave took over at 0400 and, two hours later, the wind died and the heavens opened up. The torrential downpour lasted for eight hours. Because the autopilot wasn't working, we each took our turn standing in the open cockpit, sharing the misery.

My turn at the helm started at 0800, and my foul-weather jacket was quickly soaked through. The rain, streaming over my head and shoulders, then ran down my bare legs and feet, and pooled on the sole, where it was thrust out the scuppers by the rolling of the boat. By the end of my watch, I stood huddled at the helm, dripping wet and miserable. When Dave relieved me at noon, he stopped in the shelter of the companionway and shook his head contritely, mumbling, "I'm sorry I got you into this."

Almost numb, we endured the watches because we didn't have a choice; the passage had become a grueling test of stamina. We were tired, frustrated, and not particularly pleasant to each other. We recorded only 67 nm for Day 9. When the rain eventually stopped, the sky remained cloudy, with no wind and rolling seas. At change of watch at 1600, Dave voiced the doubts that nagged at us both when he asked, "Did we make a mistake in deciding to head for the South Pacific?"

Twelve hours later, at 0400, the winds picked up from the SSE at 10–15 kt, and we motor-sailed for the last eight hours into Palmyra. We were within ten nm when we first saw the tops of palm trees on the horizon. Sailing down its south side for seven nm, we came to the spot on the chart that showed a passage through the reef. With the boat sitting precisely on the GPS coordinates, Dave turned the bow towards the lagoon.

Visibility was poor as we studied the choppy waters in front of us, and we saw no sign of a buoy or range marker. However, we did see two small islands, complete with palm trees, that weren't on our chart. Confused, we continued motoring westward along the reef, searching for the channel, and gradually drawing closer. Suddenly, with the depth gauge reading thirteen feet, we were aground. Dave backed *Windy Lady* off and sent me forward to the bow, with orders to pick a safe route through the coral. Of course we knew better, we should have hightailed it back out to deeper water, recalculated our approach and tried again—but we didn't.

It was then midday, the best time to see down into the water, and I was stunned as I studied the maze of dark, circular shadows visible in the clear, blue water. I quickly realized that the light blue water was very shallow, and that the black shadows in deeper water were coral heads. They looked to be about five feet in diameter, but it was impossible to tell how close to the surface they came; in fact, that was what we had already hit.

As there was no way through them into the lagoon, I directed Dave back to the east, but even that route wasn't clear and *Windy Lady* wasn't very maneuverable. After several bumps and another grounding, I desperately yelled, "Let's get the hell out of here!" Because we were still skirting the coral zone, Dave was able to turn out into deeper water without incident, and we continued motoring eastward.

I stayed at the bow, anxiously scanning the dark waters for some sign of the channel. Several minutes later, I spotted a dark, misshapen buoy bobbing low in the water and pointed it out to Dave. Cautiously approaching, he turned the bow toward the lagoon, and soon I could make out the far side of the channel beneath the water.

Although I kept searching, I saw no sign of a range marker. We passed by the two small islands that had caused our earlier confusion, and I later realized they appeared on our chart as reefs that covered at high tide (the chart dated back to WW II).

As we entered the wide lagoon, we could see three sailboats anchored in the distance, and a small boat speeding towards us. Within minutes, we could make out the figure of a man standing in the stern of a hard-bottomed dinghy. This was Roger, the resident caretaker, with torn shirt whipping about his upper torso and diving knife strapped to the calf of one muscular brown leg. Yelling and gesturing for Dave to follow, he proceeded to meander back and forth in front of us as he led us across the lagoon, and then showed us where to anchor.

He obviously wanted to speak to us, so Dave agreed to meet him ashore in an hour's time. We then set about tidying the boat, only to be interrupted by the arrival of a second dinghy. I was soon standing at the lifelines, talking to our visitor as if she was a long-lost friend. Maree and her husband Dave were on the Australian boat, *Byjingo*. We had dinner with them one night in Hilo, but I'd been suffering from the heat then and poor company. Now, after forty-five minutes, Dave interrupted us, saying that we needed to put the dinghy in the water.

When we went ashore, Roger toured us through the camp facilities, emphasizing that the area was off limits, except by appointment. The small house had a thatched roof but otherwise appeared to be of frame construction; it was neat and tidy, as were the surrounding grounds.

A rain catchment system provided abundant water, with huge water storage tanks, a tub (urinal) for washing clothes, and a big bathtub that sat out in an open glade. He informed us that we could have all the drinking water we needed, but could have only one bath each and wash clothes once.

Taking us into the house, he pointed out several newspaper clippings pinned to a wall. The articles, now yellowed with age, contained the gory details of two murders that had taken place some twenty years earlier, when two couples on two yachts shared the lagoon for a time. One of the men was eventually convicted of killing the other couple and chopping up their bodies. Roger, of course, told us the incident happened in the small bay where *Windy Lady* sat at anchor.

We were almost dropping from exhaustion when we returned to the boat. As Maree had invited us to dinner, we tried to nap, but were wide-awake as soon as we laid our heads down. We settled for leisurely baths in the cockpit and relaxed until it was time to row across to *Byjingo*. Significantly, we didn't talk about the passage, not even noting that we had been at sea for eleven days and sailed 1,081 nm.

That changed as soon as we joined Dave and Maree. In short order, we were pouring out all our doubts and frustrations, each of us feeding off the other. Our hosts just sat and listened, then responded with some sorely needed encouragement, providing exactly the therapy we needed. They then told us about some of the difficulties they had experienced, and *Byjingo* Dave (BD) ended by saying, "It's my opinion that most cruisers don't like passage making."

Dave and Maree were considerably younger than we were and had built their own boat before leaving from Sidney, Australia, the previous year. Their route had taken them north to Japan, across to Alaska, and down the BC coast. After wintering near the US-Canada border, they were following the same route as us across the Pacific, but had left Hilo a week earlier.

Before we climbed down into the dinghy, Maree gave me a bottle of betadine that she said was great for coral cuts, while BD advised us to check beneath our boat before going in the water. He explained that a large barracuda had been resting in the shade beneath their boat, but now appeared to be residing under ours.

So, we returned to *Windy Lady* with considerably lifted spirits and even some restored confidence. In future, we would find a similar level of companionship and generosity in other members of the cruising community. Whether it was sharing common frustrations, charts or boat parts, most yachtees were there to support one another.

The next morning, our first day in paradise, we awoke to the light patter of a rain shower on the cockpit floor above us. Taking our coffee outside, we sat and watched a small manta ray cruising beneath the rippled surface of the lagoon and listened to the soft lap of water against the hull. In the background, the roar of surf on the outer reef was constant, but varied in volume depending on the wind.

We planned to spend the day resting but the temperature rose quickly, so we rigged a tarp over the cockpit for shelter. We also lowered our new 4-hp Suzuki outboard engine onto the back of the dinghy and took a quick spin around the lagoon. I was astounded at the number of birds nesting in the thick vegetation along the water's edge. That night we returned *Byjingo's* hospitality, as the other couple had indicated they would be leaving sooner rather than later. Over a simple dinner, we shared the bottle of wine that the marina manager in Victoria gave us just before we left.

It started to rain that evening and was still raining when we awoke at 0600, so we dozed awhile longer. Just after daybreak, we heard loud, sharp knocking on the hull and a man's voice urgently calling for Dave. He scrambled from the berth, pulled on a pair of shorts, and ran up to the cockpit. I was just a minute behind and emerged to find the lagoon hidden behind a grey wall of rain. I then saw BD and Maree, wearing swimsuits, standing in their dinghy with water streaming down their heads and shoulders.

They were looking at a triangular shape poking up through the surface of the water, but I only recognized the bow of our dinghy when I saw the painter attaching it to the toe rail. The rest of the boat, including the engine, was underwater. Rain overnight had filled the boat and the weight of the engine on the stern, combined with a bit of wave action, caused it to sink.

Dave was already at the mast, releasing a halyard and passing it down to BD, who attached it to the bow. He then attempted to hold the dinghy off the lifelines, while Dave winched it clear of the water. After most of the salt water had drained out, they swung it up on deck, removed the motor, and returned it to the surface of the lagoon. With our thanks ringing in their ears, BD and Maree hurried back to the relative dryness of their own boat, leaving us staring at the waterlogged motor now hanging in the cockpit.

Dave quickly scanned the owner's manual before taking the engine ashore. He found a barrel of fresh water near the landing, where he was able to set it while he worked; meanwhile, the rain poured down heavier than ever. When the rain started to ease, the winds picked up, reminding me of another problem. Roger had been in the way when we were laying out the chain, so we hadn't been able to set the anchor. We intended to re-anchor, but hadn't done so yet, and I was sure we were drifting closer to the other boat.

Dave was gone for a couple of hours and stopped to thank *Byjingo* after he had the engine running. He then returned to *Windy Lady* and reported that they too thought the boats were getting a little close. We re-anchored at 1300, when the rain took a breather, but made three attempts before we were satisfied. The problem was that the cove was small, and we were laying out 225 feet of chain because of the deep water.

The rain then returned and it poured all night. We caught buckets of water off the small tarp that we'd hung over the cockpit, and another foot of water in the inflatable dinghy, which we'd set up on the cabin roof. We now had plenty of water for washing, so wouldn't have to use any from our tanks, or impose on Roger's hospitality.

Dave wanted to check the keel for damage arising from our encounter with the reef, but the barracuda that BD mentioned was spending a lot of time under our boat. The fish turned out to be six feet long and was trailing some thirty feet of fishing line. So, when the weather started to clear, he went ashore to talk to Roger. At first, the caretaker just shrugged his shoulders and said, "Aw, that fish isn't a problem; there's plenty of food in the lagoon."

After thinking about it for a minute, he then added, "You know, they have attacked fishermen over on Fanning Island." At that point, Dave decided to wait until the fish was gone before diving under the hull. When he did, he found no damage.

With the weather a little more settled, we made numerous dinghy trips around the lagoon, studying the thousands of birds nesting near the shoreline. Many of them were blue-billed boobies, and the babies were nearly the size of their parents but still unable to fly. At first glance, they looked like large balls of white fluff caught in the branches, and something about how they sat and watched reminded me of Daffy Duck. We saw only a few reef fish along one end, and most of the coral within the lagoon appeared to be dead. Several small manta rays regularly hunted in its waters though, and we saw two black-tipped sharks, about four feet long, swim by the boat one evening.

We spent hours walking the beaches and crisscrossed

the little island as we explored, finding many empty bottles and stumbling over cement bunkers hidden in the undergrowth. The cement foundations of a WWII gun installation dominated the white, sandy beach on the northwest corner, a grim reminder of the harsh reality with which so much of the world then lived. Palmyra, it turned out, had been a naval air station, with some 2,400 military personnel stationed there for a time.

A more recently wrecked airplane was piled up near one end of the old airstrip, and hundreds, if not thousands, of sooty terns nested alongside it. At dusk, the birds filled the air with raucous calls. One evening, when the wind died, we were inundated with bugs and I had to wonder whether any of the seabirds on the island actually fed on them. We saw hundreds of hermit crabs on the trails, a few fiddler crabs on the beach, and Roger showed us four huge coconut crabs. They were penned in a small shed in the shade of the trees and were about twelve inches across the body, with enormous claws.

Dave couldn't leave his first island paradise without husking a coconut, so one morning we set off in search of coconut palms. The first challenge was knocking one off a tree, and he then spent considerable time removing the husk, but the small nut that remained had little liquid. Later, we were able to twist a green coconut off a tree. It was full of water, as the nut hadn't formed inside yet, and provided a mild, refreshing drink.

On August 8, Dave's birthday, *Byjingo* departed for Apia, Western Samoa. We had enjoyed their company, and I watched soberly as they raised the anchor and motored out through the reef. After rocking in the swells outside, they soon disappeared to the south. That afternoon I did a large wash, and even the sheets and towels were dry in two hours. We spent the next day preparing the boat for sea.

CHAPTER 10

Passage to American Samoa

After only six days ashore, we left Palmyra on Saturday, August 10, heading south across the equator into the Southern Hemisphere. Our destination was the island of Tutuila in American Samoa, some 1300 nm away. We figured on thirteen days at sea and hoped to stay east of a rhumb line course, as that would put us in the best position to approach the harbor city of Pago Pago. If we did drift west, we could expect a difficult beat into SE trade winds at the end of the passage.

With spirits much restored, we looked forward to the crossing and even talked about sailing the entire distance. Right from the start, however, things did not go quite to plan. We'd scheduled a noon departure, thinking that underwater hazards would be most visible with the sun high in the sky. But heavy clouds hid the sun that morning, while a string of squall lines brought gusty winds and rain showers sweeping across the lagoon. We then weren't able to raise the anchor. No matter in which direction we pulled, it wouldn't budge. So we tightened up the chain, pulling the bow down into the water, and eventually, the weight of the boat broke it free.

Next, I steered *Windy Lady* out into the lagoon and turned the bow into wind, so Dave could raise the mainsail. Giving a mighty heave on the halyard, he pulled the sail halfway up the mast, and then smoothly wrapped the end of the rope around the port winch. After attaching the winch handle, he cranked it once— and suddenly, winch pieces were flying in all directions, leaving him standing with a bewildered look on his face and the handle dangling in his hand.

As parts started to roll with the gentle motion of the deck, he came out of his trance and chased them down, all the while loudly cursing the incompetence of the workers who had installed the new winches in Victoria. Fifteen minutes later, with the help of some Loctite, he had the unit re-assembled and secured to the mast. As I watched from the helm, I could only think, "Better here and now, than later, out there!"

With the mainsail up, we motored across the lagoon toward a buoy that we now knew lined up with a notch cut into the trees on shore. This was the range marker, which would take us safely out through the channel in the reef. All we had to do was keep the buoy centered on the stern and in line with the notch. At Sand Island, about halfway, the range marker was just visible in haze; I couldn't see it at all from outside the reef.

With crystal-clear water, the flat bottom of the channel was visible some twenty feet down. As we passed through, I saw the shadowy outlines of several small schools of fish, all about three feet long, and then saw the head of a sea turtle poking above the surface. It dove down about two feet and swam past, but I had a good view and figured the shell was well over two feet in length and nearly as wide.

With light winds and calm seas, we turned the bow to a heading of 190 degrees, just slightly west of due south, and continued to motor while the batteries were charging. Two miles off the island, we passed through a strong west-flowing current, where high standing waves tossed *Windy Lady* about unpleasantly for fifteen minutes. When we shut down the engine two hours later, the silence was even more remarkable than usual, with SSE winds under 5 kt and a boat speed of only two kt. We then drifted on through the afternoon, and by the start of my watch at 1600, Palmyra had disappeared over the horizon.

We continued to drift until just before midnight, when a sudden squall brought 25-kt gusts and rain showers. In the dark hours that followed, sheet lightning flashed brightly on the horizon and winds picked up to 5-10 kt. Squalls became more frequent after daybreak, and in between, winds increased to 10–15 kt. At noon, after twenty-four hours at sea (Day 1), we logged 71 nm and were ten nm east of our course line.

In the early afternoon, SSE winds strengthened to 20 kt, pushing up lumpy three-foot seas. As Otto couldn't cope, Dave now hand-steered. Near the end of his shift, he spotted a freighter off to the SSW, appearing and disappearing as *Windy Lady* rode up and down on the swells. The ship was just in radar range at eight nm and one of only a few that we saw in mid-ocean. Otto started working again when we put a second reef in the mainsail at change of watch at 1600.

By midnight, the breeze was down to 12–15 kt. Soon after, the sky cleared, revealing a canopy of bright stars, and the sea grew calmer. Before daybreak, the air had a definite chill, and with the rising of the sun, the winds veered southward. By 1000, they were on the nose at 8–10 kt, pushing us westward. For Day 2, we logged 102 nm, but when I plotted our noon position, we were twenty-five nm west of our course line.

After checking and rechecking my work, I had to tell Dave that we'd lost thirty-five nm of easting during the previous twenty-four hours. The log revealed that we'd been drifting west since early the night before, but during the last watch, we sailed seven nm south and drifted eight nm west. Apparently, while we concentrated on keeping the bow pointed south, strong west-flowing currents just carried the boat sideways.

Hoping to regain some of the easting we'd lost, Dave started the engine at 1300, and *Windy Lady* pushed smoothly forward under a clear, sunny sky. The winds then eased a bit more, dropping down to 5–7 kt, and the ocean flattened out, with barely a swell on its surface. The temperature soared, however, and soon the only tolerable spots on the boat were the rear cockpit seats, which caught a surprisingly cool breeze. Before long, it was ninety-five degrees inside the cabin and Dave unbolted the galley hatch, leaving it wide open for over thirty hours. The berths were then as comfortable as any bed ashore, but we were only able to sleep with the small 12-volt fans directing air onto our heads.

Although we suffered with the daytime heat, the night watches were extraordinary. First, I watched the sun disappear into the sea, but didn't see the green flash that Gordon had told us to look for. (He did say that we might be sitting too low in the water.) I then searched the sky for pinpricks of light as the stars appeared. With no haze on the horizon and not a cloud anywhere, the night sky soon extended all the way down to the sea in every direction. More stars were visible than I'd ever before seen, with the brightest reflected on the quiet waters around me. I even identified Betelgeuse as it climbed over the horizon in the east.

Most remarkable, however, and a sight we've never forgotten, was the Perseid Meteor Shower. With a dark moon, we watched raptly for hours as hundreds of meteors streaked amongst the stars. At midnight, when I climbed up the companionway, I saw a bright light in the sky behind Dave's shoulder. He thought it was lightning but I saw a fireball flare to half the size of a full moon and then vanish. We continued to watch even into a second night.

After motoring for thirty-nine hours, we started sailing again at 0400, with a SSE breeze of 8–12 kt. Although within fifty nm of the equator, Dave was looking for a sweater before daybreak. The morning that followed was spectacular, with a clear, cloudless sky, warm sunshine, and cool breeze. Within a few hours, SE winds of 10-20 kt were pushing up three-foot seas and the views around me were extraordinary. Awestruck, I could only marvel at the wondrous good fortune that had brought me there.

In every direction, inky-blue ocean waves rolled ceaselessly towards the west. White caps sparkled here and there on the surface, while the occasional shower of white spray flew up off the bow as *Windy Lady* cut smoothly through the waves. Overhead, not a single blemish marred the huge blue dome of the sky, with the deep, dark center at the zenith blending smoothly into a whitish-blue hue around the horizon. Apart from a few flying fish and a couple of birds, we could have been alone on the planet.

Shortly after midday, the winds increased to 15–20 kt and backed to the ESE. *Windy Lady* was then on a beam reach and flew across the ocean for hours. Not willing to give up one moment, I spent the afternoon in the cockpit with Dave. We crossed the equator at 1625, standing in front of the GPS and watching as it changed to read 0 degrees of latitude. As the sun dropped into the sea and the stars came out, the waves grew a little higher—but our magic carpet ride continued for nearly twelve hours.

By midnight, winds were gusting from 15–25 kt and Otto couldn't cope, so we put a second reef in the mainsail. When that didn't help, I was back to hand steering. I continued to tweak sails and vane settings for the next three hours, and suddenly, Otto surged back to life and worked perfectly for the next five hours. By noon, winds were at 15–20 kt from the SE, and seas had grown to six feet. We recorded 97, 106, and 120 nm on Days 3, 4, and 5 and more than recovered the easting we'd lost.

The winds were down to 12 kt when I came on watch at 1600; by midnight, they were below 10 kt and had backed to ESE. A few hours later, the GPS revealed that the currents were still with us. I first noticed a boat speed of 3.7 kt and a course of 148 degrees; only minutes later, the boat speed was 5.0 kt, with a course of 227 degrees. All the while, the compass was pinned on 180 degrees.

When the dawn broke, clouds were building on the horizon, and fast-moving squall lines swept down on *Windy Lady* throughout the day. The squalls brought gusts of 25 kt in the forenoon and 30 kt in the afternoon; in between, winds were light and variable and we had difficulty keeping the sails filled. The storm system disappeared with the setting sun, and winds remained light overnight.

By morning, the breeze barely registered at 0–5 kt, and *Windy Lady* rocked uncomfortably in three-foot swells. The temperature again soared, and we sweltered inside the cabin and roasted in the cockpit, covering up with long-sleeved cotton shirts and wide-brimmed straw hats when on watch. About midday, the swells grew to six feet and the boat rocked mercilessly. We logged 102 and 98 nm for Days 6 and 7, at which point we were about halfway to Pago Pago.

As the afternoon dragged on, the air finally stirred out of the ESE. By 1600, winds were up to 10–15 kt, and they continued to strengthen during the early part of the night. Winds gusted from 15–25 kt throughout my midnight watch, and I kept busy easing the sails when the gusts hit and hardening them when they eased. During the twelve hours ending at 0400, we averaged 6.1 kt, with a top speed of 7.0 kt.

Although the winds eased a little after sunrise, seas grew much rougher, so we slowed the boat by putting a second reef in the main and partially furling the headsail. For Day 8, we recorded 136 nm and picked up ten nm of easting. Shortly after midday, the winds backed to the east and settled at 10-20 kt. They were then abaft the beam, and rough six-foot seas now broke alongside. The occasional wall of spray flew up over the cockpit, and before long, seats and deck were feeling grimy.

That didn't stop us from sitting in the starboard seat for nearly an hour, captivated by the sight of some fifty dolphins playing in the oncoming swells. Mostly, they surfed, staying just below the breaking tops and riding the waves in toward us. That reminded me of a fish I'd seen the previous day, swimming off the stern on a following sea. I assumed it was a shark, as it was white, very large, and instantly disappeared.

The winds strengthened next morning, veering to ESE, and seas grew higher. By noon, winds were gusting from 15–25 kt and seas were about eight feet. The currents now appeared to be in our favor, as the GPS registered a half mile more than the Signet during the following watch. The winds then alternated between 10–15 and 15–20 kt at the end of each of the next six watches. Towards the end of that time, the GPS recorded a boat speed of 11.1 kt, which I simply couldn't believe. Otto performed well and we logged 125 and 123 nm for Days 9 and 10.

For the next few hours, winds seemed to be easing,

but actually, we were experiencing the lull before the storm, and the storm was the one I'd been dreading ever since leaving home. Shortly after I came on watch at 1600, winds increased to 15–25 kt; before my shift was over, they were at 30 kt. I was quite happy to turn the helm over to Dave, but then laid awake for the next four hours, with my berth jerking and twisting beneath me.

When he rousted me out at midnight, winds were hitting 35 kt, and lumpy ten-foot seas were causing the boat to lurch about wildly. Caught unprepared as I crawled from my berth, I whacked my head painfully against a bulkhead. I made sure to keep a good grip on the grab rails after that, and made my way forward to the settee amidships, where I'd left my clothes and rain gear. I found it easier and safer to get dressed there in heavy weather.

My heart sank as I listened to the storm raging outside, and I glanced over at Dave, who was hanging onto the chart table, trying to fill in the journey log. Even in the dim light of the 12-volt system, I could see water dripping off his foul weather gear onto the sole. I then watched, hypnotized, as the small pool flowed back and forth over the narrow teak boards. The next thing I knew, I was dressed and groping my way up the companionway.

Stepping out into a pitch-black night, I inched my way around the helm, conscious of the wind tugging at my jacket and rain stinging my face, while cold water sloshed over my bare feet. I then sank down onto the rear starboard seat and managed to respond to Dave as he adjusted the sails, furling in the headsail a little and easing out the main, so the end of the boom stretched well out over the waves. Before going below, he paused and watched as I braced my legs against the end of the forward seat, then asked, "Are you going to be all right?"

Looking up, I saw him outlined against the weak light from the cabin, bobbing about as the boat rolled and pitched. Knowing I didn't have a choice, I nodded my head and managed to squeak, "Yes, of course," but the words were lost in the roar of the storm. After starting down the companionway, he stopped and boarded up the entrance for the first time; moments later, the cabin lights went out, and I was alone.

Out of habit, I looked at the wind gauge when I first sat down. It read 35 kt, which so traumatized me that I never looked again. I then saw that the storm was overpowering Otto, so grabbed the helm and held the bow on a heading of 180 degrees. Braced in my seat, with my neck craned out to see the compass and my hand on the wheel, I then rode out the twists and turns of the boat with rain streaming down my face and jacket.

I kept my eyes glued to the face of the compass as though it was a lifeline to reality, and time ceased to exist. At some point, I became aware of feeling cold but only later, when I felt the helm respond to the self-steering, did I become conscious that I was shivering and soaked through to the skin. When I realized that Otto was holding the bow on course for minutes at a time, my confidence started to return and I dashed below for a sweater— and saw that two hours had gone by.

When I climbed back into the cockpit, I saw some definition in the sky behind and hope surged through me that maybe the worst was over. I took a seat on the port side and made a grab loop by knotting a short piece of rope that was hanging on the push pit railing. Holding onto it for security, I dared to look around and, over the cabin roof, saw low, dark clouds scudding across a black sky; the wind gauge was then reading from 30–35 kt. With *Windy Lady* rolling, turning, and pitching, the mast was swaying down on one side, then the other, and the strobe light flashing on the masthead revealed angry-looking waves breaking all around.

The winds then veered twenty degrees and eased a bit, but the seas grew higher and the cockpit wetter. Before the end of my watch, breaking waves dumped the equivalent of a bucket of water over me at least six times, and I couldn't count the number of times I took a glassful in the face. As the last hour dragged on, I felt the occasional gust of warm air on my cheek and wondered where it came from. We made twenty-three nm during that watch.

When I went below at 0400, settee cushions were lying on the sole, as were charts and pencils off the chart table and a basket from the galley table. Water had come in the forward hatch and was sloshing about on the sole in the head, but the cabin windows were dry, as Dave had re-sealed them in Palmyra. I wiped up the water we had tracked below and found myself wondering why we hadn't put a third reef in the mainsail.

I now checked the port berth and found the oars lying in the middle of the mattress. We'd stowed them alongside, so I tried to re-position them but they simply returned with the next roll of the boat. As I struggled to secure them, my hand touched a soggy corner of the mattress and I realized, with a sinking feeling, that water was seeping in from a nearby porthole. So, grabbing my pillows, I headed for the starboard berth, where I slept fitfully for a couple of hours.

Dave made another twenty-two nm during his watch, but conditions remained very wet. His comment in the journey log at 0800 reads, "Oh, what a night! Wild and wet!" Winds had now dropped to 20–30 kt from the SSE, but seas were unbelievably rough. The waves looked like small mountains, steep, close together, and twelve to fifteen feet high.

I felt a strange mix of terror and delight as I stood my watch that morning, with waves building, cresting overhead, and then crashing down somewhere along the length of the boat. Spray flew in all directions, dousing the deck and cockpit with salt water. Now and again, a large swell broke amidships, washing the keel to starboard, and *Windy Lady* then heeled sharply but quickly bobbed upright. Of more concern were those occasions when she pitched almost straight up the face of a wave and then dropped down the backside, burying the bow in the bottom of the next one.

We recorded 127 nm for Day 11, and the clear morning sky now gave way to afternoon clouds. The seas began to ease and, by 1600, were down to ten feet. An hour later, I sensed a change in the wind, just momentarily at first, but gradually the lulls became longer and more frequent. At 2000, winds were at 15–30 kt; at midnight, 15–25 kt;

and at 0400, 15–20 kt. During that time, the seas went from ten feet to three.

Next morning, I awoke to a beautiful, clear, sunny day, with 18 kt of wind and friendly seas; at 0815, we caught our first glimpse of the misty outline of the island of Tutuila. Relaxing in warm sunshine, we watched the surging sea break on rocks and reefs, while the ridges and ravines of the island's rugged eastern slopes revealed themselves. With quieter seas, we made good time, averaging 6.1 knots over three successive watches and logged 130 nm on Day 12.

We were required to contact the Harbor Master in Pago Pago before we arrived, so Dave called in on the VHF radio when we were four hours out. No one ever responded, although he tried every fifteen to twenty minutes. We started the engine just outside the harbor, dropping the sails and raising our yellow quarantine flag, then headed in through a broad reef, with surf tumbling tumultuously on both sides.

As we made our way to the customs dock, a man standing on the foredeck of another sailboat pointed in front of us and bellowed, "Reef! Reef!" After almost having joint heart attacks, we double-checked the chart and realized there was no danger. On Thursday, August 22, our thirteenth day at sea, we tied up at 1430 local time, having sailed 1,376 nm.

Dave was careful to step off the boat only to handle the mooring lines, but we soon caught the attention of a rather irritated off-duty policeman, who waved me over from where I was working at the bow and snapped, "Go get your old man!" Although Dave explained that we'd been unable to raise the Harbor Master, the officer insisted that we call again and again over the next few hours. He then disappeared for a while but returned wearing his uniform and stayed until dark. Meanwhile, another boat rafted up beside us, and a parade of people began walking across our decks.

The Harbor Master apparently had gone fishing that day, but answered his radio when we called the next morning. Officials soon arrived and cleared us in, and I then hosed salt off deck and hull while Dave took charge of a fuel delivery. We let out 200 feet of chain when we dropped the anchor in the inlet, as we'd been warned that 20-kt winds were frequent and the holding not very good. After waiting the usual hour, we put the dinghy in the water and headed towards a small jetty near the head of the bay.

I was eager to go ashore and stretch my legs, but lost some of my enthusiasm when hit by a constant spray of salt water from wind waves striking against the side of the dinghy. I became even less interested when we followed a road from the jetty and found ourselves in amongst dilapidated buildings, with piles of trash and scattered debris everywhere. Styrofoam food containers and drink cans were particularly plentiful near several 'No Littering' signs. A few scrawny chickens and mangy-looking dogs poked through the mess, and the only person we saw was busy pitching rocks at one of the dogs.

It was not an auspicious start to our visit, but things did improve over the next few days. First, trade winds eased, so wind waves and salt spray were no longer a problem. Although we did have a few rain showers, freshwater dried and didn't leave me feeling damp and uncomfortable. We then located a wet market that was within easy walking distance and were happy to have fresh bananas, coconuts and papayas again. Despite the fact that there was no obvious gathering spot for crews on boats anchored in the inlet, we also met a few cruisers when we were out and about.

Trade winds then returned with a vengeance, tearing through the anchorage at 30 kt, so we stayed onboard for two days straight. Although we didn't notice any boats dragging their anchors, I did see folks getting very wet in their dinghies, so started wearing my poncho when we did go ashore, which more or less kept me dry. The water in the inlet was actually quite dirty and left an oily scum on *Windy Lady's* waterline.

Two tuna canneries and a power plant occupied the narrow beachfront running along the north side of the inlet; the canneries proved smelly when the wind blew and the power plant noisy when it didn't. Countless plastic shopping bags also dotted the surface, and although we were careful, we managed to wrap a bag around the prop on the outboard engine. Apparently, the bags ended up on the bottom of the inlet, where they could cause problems when anchoring.

We started searching for swimming/snorkeling areas and stopped at a government building marked 'Marine', which turned out to be a fisheries office. The young woman at the counter was very helpful, saying that she too liked to snorkel. She first warned that the harbor was polluted, as was the coast near the tank farm. She then explained that currents were strong in some areas, while in others, there wasn't enough water over the reefs for swimming. She did suggest a spot that had been dynamited when the airport was built, saying that access was across the end of the runway. But when she explained that it was only necessary to follow the trail of garbage, we decided to give it a pass.

By then we'd discovered the dozens of small, wildly painted buses that made up the city's transportation system. The vehicles were highly personalized, with rugs, ornaments and plastic flowers around the driver's seat; each also had an elaborate sound system (one even had two music video screens). Most buses held about twenty passengers and the fare was 25¢ or 50¢, depending on how far you were going. We never waited more than five minutes for a ride and observed the curious habit of a few young riders, who carried a 25-cent coin tucked in their ear. We also noticed many small pickup trucks on the roads, all of them with two-to-eight passengers sitting in the back.

The Samoan people we met were soft-spoken and friendly, but very large. It seemed that at some point, normal sized children suddenly exploded into immense adults. The men were big and tall, with large hands and feet, and the women were the same. We were told that diabetes was a problem, apparently made worse as traditional diets were replaced with modern food, particularly fast foods and coke.

There were newer homes and larger shopping areas

around the point, south of the inlet, and we began to wonder how people earned a living. Other than the fish canneries and power plant, there didn't seem to be much work available, but everyone seemed to have money for fast foods and laundromats. After visiting the local hospital, where Dave paid a $2 fee to have a rash treated on his foot, we concluded that the American government probably footed the bill one way or another.

We stayed in American Samoa only long enough to rest up, about the same length of time as the passage from Palmyra. Oddly enough, I had no concerns about going back to sea. In fact, I recorded a rather surprising conversation in my journal shortly after we arrived. A fellow cruiser had asked about our crossing and I responded, "It was just about perfect." I then explained, "We had mostly good weather and smooth sailing, a couple of exciting days towards the end, and then an easy entry into harbor."

When I first reread my journal, I had trouble reconciling those words to my description of the night in the storm. My initial reaction was to wonder whether the euphoria of making landfall could have caused temporary amnesia. But my account of the following day reveals that, even then, I had put the paralyzing fear of the night behind me. Certainly, in the years that followed, there were other storms, even worse storms, but I never again experienced anything similar to that night. It seems likely that my imagination was again the problem, and that having lived through my nightmare, the fear simply dissipated.

CHAPTER 11

Vava'u Islands in Tonga

We left Pago Pago on Tuesday, September 3, motoring out through the reef just before noon. Our route would now take us southwest to the village of Neiafu in the Vava'u Islands, the Kingdom of Tonga. The distance was 325 nm and we expected a quick passage, with winds on the beam all the way.

We raised the sails as soon as we were clear of the entrance, setting off under a partly overcast sky with east winds at 10–15 kt. Happy to be heading out to sea once more, we spent the afternoon together in the cockpit. Dave took the first watch and when I relieved him at 1600, we were clear of the island, and ESE winds of 15–25 kt were pushing up lumpy, three-foot seas. Four hours later, the gusts were hitting 30 kt, and I'd eased out the main and partially furled the jib.

At midnight, SE winds were steady at 15–20 kt, and

stars were peeping through the clouds. The winds then moved around a bit but Otto performed flawlessly. By daybreak, winds were again in the SE and conditions were perfect during my morning watch. Deep-blue ocean swells about three feet high rolled silently past, and a few fluffy, white clouds floated in a bright sky. Lulled by the soft rustle of water curling off the bow, I contently soaked up the sunshine, while enjoying the coolness in the breeze. At noon, after twenty-four hours (Day 1), we logged 129 nm.

The winds then eased a little, down to 10–15 kt, with seas still about three feet. Because the passage was short and the sailing enjoyable, I kept Dave company again that afternoon. The sun grew a little warm for a while, but as it dropped in the west, the air cooled off quickly. By 1800, about half-an-hour before sunset, I was digging out a sweater. With clear skies, the stars were brilliant for the first part of the night. About 0100, a half-moon rose up out of the sea, its light creating a pathway across the quiet waters behind us. As it climbed higher, shadows appeared in the cockpit, and it was just a very pretty night.

The breeze started to ease after midnight and by 0400 was down to 6–8 kt. It dropped to 3–8 kt during the next watch and we made only eleven nm. The winds returned about midmorning, increasing to 10–15 kt, and my watch was then much like the day before, with a few scattered clouds, quiet rolling seas, and comfortable temperatures. We logged 107 nm on Day 2.

Within a few hours, the winds again eased and stayed below 10 kt for almost ten hours. They picked up again after midnight and, by 0400, were gusting from 10–20 kt, pushing up choppy seas. At first light, about 0630, the islands of Vava'u were visible to the southwest. We ran into strong currents near the first headland, so Dave started the engine and we dropped the sails. After passing a couple of small islands on the way in, we turned north into the channel leading to Neiafu.

About then, the VHF radio came alive, as "*Windy Lady, Windy Lady*," came over the airwaves. The voice belonged to Dave on *Byjingo;* they were on their way to one of the outer anchorages. In a brief conversation, we learned that they'd had a difficult beat coming down from Western Samoa, with headwinds of 25 kt the first day and 20 kt thereafter.

The channel now narrowed and curved around the base of a distinctive-looking hill, so we kept busy looking for buoys and following the range marker on the hillside. It then opened into a large, sheltered lagoon that was probably one of the safest harbors in the South Pacific. The village of Neiafu was located along one side, with Customs situated at a large commercial wharf near the entrance.

Concerned about gusty winds and currents, Dave kept the boat speed up as we approached the wharf. I was standing on the foredeck, some twenty-five feet in front of him, and immediately saw that the solid wooden structure was the same height as *Windy Lady's* stanchions and lifelines. We had only tied up at floating docks before and I grew nervous; we seemed to be coming in very fast.

As we came alongside, a group of ten young men

appeared out of nowhere and threw themselves against the toe rail, straining to hold *Windy Lady* off while her fenders were squashed flat. Happily, the boat never suffered a scratch. The men had been hidden from view by containers piled on the dock, and I later suspected they had been drinking or smoking, or whatever similar activity these young men did.

When we asked them about Customs, we learned that it was Saturday, not Friday, and that the office was closed for the weekend. Unbeknownst to us, we had crossed the International Dateline and lost a day. We had assumed the line followed 180 degrees of longitude, but it doesn't. It actually jogs around a number of Pacific islands, including them in the eastern hemisphere. Maybe that is why Tonga calls itself "the Place Where Time Begins".

With our yellow Q-flag up, we motored away from the dock and picked up a mooring buoy. Contacting the owner on VHF radio, we arranged to use it for two nights. Loud music drifted across the water for a couple of hours that evening, coming from a tourist bar we assumed. Sunday morning, we awoke to the pealing of church bells and the day that followed was sunny, warm and peaceful. We spent most of it sitting out in the cockpit, where the sounds of "sweet song" occasionally drifted to our ears. The singing came from different locations, almost as if various church choirs were in competition.

The enforced inactivity wasn't that bad, as we caught up on our sleep and talked about how we would spend the time during the two to three months we would be in the islands. According to our cruising guide, most Tongans in Vava'u were religious, living traditional lifestyles. It was against the law to swim on Sundays, although apparently that didn't apply to tourists, as long as they didn't swim in close proximity to villages. In fact, we seldom saw local people swimming; when we did, they went in fully clothed.

We were up early on Monday morning, preparing to move the boat back to the wharf. The day was grey and showery, which pretty much reflected my mood, as I hated the thought of going anywhere near it. When I looked out and saw two Tongan Navy patrol vessels tied up alongside, I had a sudden surge of hope that maybe it wouldn't be necessary. That thought didn't last long, however, as both boats left the dock and anchored out in the lagoon.

A few minutes later, the bow of a huge American warship blocked our view of the hill that marked the entrance. The *USS Frederick* (LST 1184) came into view slowly, dwarfing the hills and the lagoon itself, then carefully turned around and approached the wharf. As it drew near, the crew threw lines down to a small powerboat that delivered them to men onshore; once the mooring lines were secured, the massive vessel was pulled in. The warship was some 550 feet long and the wharf only 250 feet, so the stern stuck out a bit on one end, the bow far out on the other.

While this was going on, Dave dinghied ashore to find out what we were supposed to do. A Customs officer told him to wait until the Americans finished docking and then bring *Windy Lady* alongside the end of the wharf.

An hour later, when that didn't look possible, he went ashore again. The officers now said they would come out to the boat.

I was still watching the activity ashore when our dinghy appeared around the stern of the ship; I couldn't believe my eyes. Talk about three men in a tub! The Tongans were big men, bigger than Dave, and their combined weight had the boat, including the bow, sitting low in the water. With barely an inch of freeboard all around, the dinghy mushed through the water, looking like a small wave could sink it. I immediately glanced around quickly, making sure there was no traffic nearby.

Only slightly damp, the two officers, Customs and Agriculture, were on board for half an hour, looking around, asking questions, and filling in forms. When Dave took them ashore, he returned with the Immigration Officer, who barely stepped aboard before demanding a soft drink. With no refrigeration, we could only offer him warm tang, which he unhappily accepted; he cleared us in anyway. After taking him ashore just before noon, we removed the Q-flag and anchored out in the lagoon.

We stayed in Neiafu for nine days on this occasion, although we returned regularly during our time in the islands. Daytime temperatures were very comfortable, and despite the threat of rain, we spent hours walking through the village. Small wooden houses nestled under the trees on large, residential lots, while pigs of all sizes and varieties roamed freely, as did chickens and dogs. The dogs were much healthier-looking than the ones we'd seen in Pago Pago. Both yards and streets were reasonably clean and vehicle traffic almost non-existent, although the occasional pickup truck did go roaring past.

The first afternoon, we followed a narrow, dirt road through the shopping district, which stretched about three blocks. Most of the small shops were located in wooden buildings on the waterfront side of the road and included a couple of eating-places, a bar, several small grocery stores, and two supermarkets. The stores carried a limited supply of canned goods and staples such as potatoes, onions, rice and flour; some also carried frozen meat, mostly lamb. Spam and tinned corned beef appeared popular but prices were steep, with a tin of spam costing $5.50.

The bank, when we came to it, didn't have an ATM, so we asked for a cash advance on master card. It turned out the telephone lines to Nukualofa, the country's capital city, were down, so we had to return a couple of hours later. We then located the wet market, as well as several businesses that catered to the cruising community. One of these was a bar overlooking the lagoon, which turned out to be a favorite haunt for some sailors. Five minutes after dropping the anchor, they could be ashore, sipping on cold beer while keeping an eye on their boat.

Another business offered boat repair services, rented mooring buoys at $5/night, and sponsored a weekly potluck barbeque for yachtees. We looked forward to attending, as this would be our first opportunity to meet members of the current year's cruising fleet. I made a pot of homemade baked beans to take along, and Dave tracked down a local entrepreneur by the name of Hans and bought a couple of very expensive steaks. (Hans brought frozen meat up from the capital city in his boat.)

Over forty people attended the barbeque, so we weren't disappointed. We learned that boating was the main tourist attraction in Vava'u, but business was slow, as it apparently was throughout the South Pacific. Upwards of fifty sailboats based their operations in Neiafu during the cruising season, while two bareboat charter operators catered to sailors flying in for a week or two at a time. That year, many of the cruisers were Americans, some having originated on the east coast, while others sailed from the west coast after spending a winter in Mexico.

With thirty anchorages within a few hours' sail, the boats moved around freely. Some sites had good swimming beaches, others were close to snorkeling destinations, and a few provided a weekly feast featuring local foods and entertainment. In the last two years, restaurants/bars had also opened at two sites, both operated by white men with Tongan wives.

The topic that dominated conversation that night, however, concerned a British couple and their sailboat. While Dave and I heard different details, the general theme was the same. The couple had been together for nineteen years but weren't married; during that time, both had worked to maintain the boat and their lifestyle. Trouble then came to paradise in the form of a temporary female crewmember, and the man had recently announced that he was going to marry her. As a result, friends of the first woman were lobbying intensely on her behalf.

The three people concerned were present at the

barbeque, and much whispered conversation flowed around them. They were all living on the boat, as neither party was willing to give up possession; after all, where would anyone go? In the days that followed, the lobbying even spilled over to the morning broadcast on VHF radio. We didn't follow the ins and outs of the situation, so only learned how it ended when we returned to the islands three years later. The first wife had the boat and still sailed in the islands; the new couple had married, bought another sailboat and moved on.

This event caused some soul-searching amongst yachtees, and not necessarily because of concerns about romantic third party involvements. It was simply a reminder that the laws offshore did not provide the same protection as those at home. Our only concern was for *Windy Lady*, so we double-checked to make sure that I would be able to take her out of the country legally, if anything happened to Dave. Nonetheless, a few months later, we did hear reports of two marriages that took place in New Zealand.

While in the islands, cruisers kept in touch using VHF radio, and a couple running a beach restaurant on Hunga Island provided a report every morning on local conditions. As well as the weather forecast and tide schedule, they included items of interest such as the current week's feast. One morning, we heard that the American sailors and marines on the *USS Frederick* were in Tonga for exercises with their local counterparts, and that the ship would be open for tours later in the week.

At the appropriate time, we made our way over to

the ship and a young sailor guided us around the deck; he explained that it was an LST, a tank landing ship. Taking us up to the bow, he described how it would be driven up on a beach, large doors opened at the front, and equipment driven off. As I was trying to get my mind around this concept, he mentioned that they had stopped at a small island on their way down from Pearl Harbor and picked up an injured sailor off a yacht.

His curiosity immediately aroused, Dave asked, "Was the island Palmyra?"

The sailor shrugged and said, "I don't know what island it was, but we steamed back a couple of hundred miles." He then explained, "We sent in a boat with a doctor, and that was lucky, as he was only on board because of the contingent of marines we carried. We then took the man over to Christmas Island and put him on a US Coastguard Hercules that flew down from Honolulu."

He then recollected something else and added, "The man was apparently in pretty bad shape; he'd been swimming and was attacked by a barracuda."

Of course, we just knew the island was Palmyra and the barracuda the one that had rested beneath our boat. It wasn't long before yachtees on HAM networks had confirmed the details. A Kiwi boat had stopped at Palmyra on its way back to New Zealand after the Trans-Pac yacht race. A crewmember jumped into the water and was struck by a barracuda, severing an artery in his thigh. We concluded the fish had probably been resting under the boat, just as it had under ours, and lashed out when the water erupted beside it. At that point, Dave remembered how close he'd been to going into the water with that fish and said he could feel his own thigh aching.

A few days later, we heard band music wafting on the breeze and rushed up to the cockpit to see what was happening. The US Marines, along with a few sailors, were parading through town, partly I guessed, to thank the Tongans for their hospitality. It took about fifteen minutes for the band and about 100 men to march down the main street—in the rain! We got quite wet as we listened to the music but were rewarded by a rather misty glimpse of the men marching along the cliff road. As I stood there, I thought of the sailor in Palmyra who probably owed them his life. It seemed I had gained a new perspective on the US military.

We were now ready for some swimming and snorkeling, so were planning to visit some of the other islands. As we couldn't obtain a local cruising guide, I pored over our charts, following routes through the reefs as I looked for protected anchorages. Coral seemed to grow everywhere behind the barrier reef, so that even in what appeared to be open water, it was necessary to follow channels and passes.

We chose a sunny morning for our first trip but didn't go far, just around the corner to Port Morelle, anchorage #7. Wind and water were quiet as we approached the small, sheltered bay, and I could see the masts of three sailboats etched against the palm trees. While Dave circled *Windy Lady* around, checking for depth, I stood at the bow, studying the yachts floating serenely on their tethers and the waves washing softly up on white sands.

Wanting to stay well clear of the coral heads, which extended out to a depth of thirty feet, we dropped the hook in sixty feet of water. Our CQR anchor then slid across the smooth, sandy bottom, and we tried three times before it dug in. Wanting to see how well it was hooked, we put on snorkeling gear and pulled ourselves down the chain until we could see the anchor buried in the sand. We then swam ashore and were dismayed to see anchors and piles of line/chain sitting on the sand beneath the other boats. The crews had made no effort to set their anchors, and that turned out to be standard operating procedure for quite a few cruisers.

Port Morelle was a convenient stopping place for boats coming and going from Neiafu, and the small bay could become very crowded. Cruisers often arrived late and left early, and I now kept a close eye on anyone anchoring nearby. Still, a late arrival would invariably sneak in just before dark, trying to find a spot amongst fifteen or more boats, and often ended up a little too close for my comfort. Fortunately, while we were there, the winds were only strong enough to lift the canopy and rattle the lifelines.

We started using the inflatable dinghy again, as it made a much better swimming platform than the hard dinghy. I swam every day, trying to get used to my snorkeling gear, but found the water cool and could stay in for no more than thirty minutes. Palm trees shaded the narrow strip of coral reef extending around the shoreline, so visibility was good only at low tide with the sun high in the sky. Still, many schools of small fish darted about, including some that were very tiny and electric blue, while several larger, white fish poked their beaks into the sand.

It was here that we again spent an evening with Dave and Maree on *Byjingo*. Our plans for dinner went awry, however, when the grill plates for the barbeque disappeared into sixty feet of water. Still, we were able to cook dinner on the propane stove in the galley. As the evening was warm and pleasant, we sat outside talking long after the stars appeared. When I asked, Maree was very happy to point out the constellation known as the Southern Cross.

Losing items overboard was a regular occurrence when you lived on a boat. Over the years, we lost tools, replacement parts and keys. What was most frustrating was that the item seemed to float through the air in slow motion, so you clearly understood what was happening but were powerless to stop it. That night, Dave clamped the barbeque onto the starboard rail before placing the grill plates inside. When he turned away, he heard a grating sound and turned back just in time to see the whole thing slip backwards. He then watched the plates fly through the air and slowly settle to the bottom of the bay.

CHAPTER 12

Hunga Island

After spending a few days in Port Morelle, we returned to Neiafu and replenished our supply of fresh fruits and vegetables. The next morning, with the sun peeking through the clouds, we set sail for Hunga Island on the leeward side of the archipelago. With steady 20-kt winds filling the sails, *Windy Lady* raced across smooth, sparkling blue waters for over an hour. We noticed a slight swell when we turned down the outside of the island, but then lost the wind and had to finish the trip under power.

The narrow passage into the lagoon was rather surprising, as it cut through a low ridge that I thought was part of the island, but turned out to be the outer perimeter of the lagoon. A rocky outcropping in mid-channel restricted the passage even more, and then swells tried to push *Windy Lady* through sideways. I kept watch at the bow, searching for the buoys that would lead us through the shallows on the other side. I always worried about the six feet of hull beneath the water, so was greatly relieved when I spotted the first one.

Hunga lagoon was quite large and almost rectangular, with a village located on the northeast corner. A small restaurant, Club Hunga, sat on a white, sandy beach on the northwest corner. As we entered the lagoon, we could see the masts of three sailboats anchored near two islands south of the village, so we motored over to join them. The anchor hooked nicely on our first attempt and the boat settled in place, with our small Canadian flag fluttering on the stern.

We spent the afternoon quietly but noticed a number of small powerboats and outrigger canoes coming and going from the village. Later, as we prepared dinner in the galley, we were surprised to hear soft tapping on the hull. Going up to the cockpit, I saw a dignified, older Tongan man sitting in a dugout canoe alongside. After we had greeted each other, he admired our boat and I admired his canoe, noting that the curved sticks attaching the outrigger pole appeared to be tied on with fishing line.

Speaking quietly, he then explained, "I live in the village and my name is Vaha. I've come to invite you to my church on Sunday and then to my home afterwards for a Tongan dinner."

I promptly called down for Dave, whose first question was, "Real Tongan food?"

Upon being reassured, he enthusiastically accepted and asked if we could possibly buy a few bananas or coconuts. When that too was confirmed, the men agreed to meet at the village jetty at 1000 the following morning.

Intrigued by the invitation, we stood and watched the Tongan paddle away, not yet aware of the unique experience awaiting us.

Promptly at 1000 next morning, we arrived at the jetty, which turned out to be just a pile of rocks and dirt dumped out into the water. As there was nowhere to tie up, I scrambled out onto the bank, and Dave then anchored the dinghy and waded ashore through the mud. Vaha appeared some ten minutes later, carrying a large basket, loosely woven from leaves, containing plantains, coconuts, a sweet potato and a stock of green bananas. After some discussion about payment, he stated quietly but firmly that he was a friend and this was a gift. Somewhat taken aback, as this was not what we'd intended, we accepted the basket as graciously as we could.

Vaha then surprised us by asking, "Do you want to see my plantation?" Of course we did, so he led us up the hill and through the village. As we climbed, I noticed two low fences, made of sticks, running out into the lagoon on either side of the jetty. In response to my question, he explained, "The fence surrounds the village and keeps the pigs inside and away from the crops."

He then added, "There are seventy families in the village, about 380 people. We have five churches, three Tongan (Wesleyan), one Mormon and one Methodist." When asked, "Why three Tongan?" He responded, "Some people wanted to be on their own."

We crossed a large, closely-cropped grassy meadow that formed a common area between the houses, and saw a few chickens, goats and horses roaming freely, as well as several pigs rooting about under the trees. The area seemed remarkably clean, even to the point that the ground beneath the larger trees had been swept. The small homes were mostly of frame construction, with low fences of barbwire, sticks and corrugated steel around them, while the sheds and outbuildings all had thatched roofs and walls of corrugated steel.

Each house was equipped with a small solar panel that provided power for a few lights, and a large, cement cistern that caught rainwater off the metal roof. Curiously, Dave asked, "Have you ever had a drought?"

Vaha smiled and answered, "No, we always have lots of water." He then continued, "But we have no power tools, all the work in the village is done by hand. We use the horses to carry supplies up from the beach and to move crops down from the plantations." However, I did notice a generator near the jetty and heard someone using a power tool on the bottom of an upturned boat.

We saw only a few people as we walked through the village, and Vaha politely greeted each one. Clothes were hanging on lines, drying in the warm sunshine, and a woman was peeling bark off a sapling, making tapa cloth, we were told. Four young girls, maybe five or six years old, then spotted us and ran over, a group of slim, dark-haired, dark-skinned, brown-eyed little ragamuffins. With bare feet and torn, dirty dresses that hung below the knee, they eagerly asked me my name, wanted to know where we were from, and then volunteered their own names. When I hurried off to catch up with the men, two of the girls grabbed my hands, trying to hold me back, but let go when we got closer and quickly disappeared from view.

At the back fence, Vaha opened a rough gate made of sticks and corrugated steel, and we set off down a narrow path that wound through the trees. After a short walk, we reached his plantation, which was just an opening in the forest beside the trail. He pointed out various plants as we came to them: banana, coconut, papaya, sweet potato, yam, breadfruit, cassava (from which comes tapioca), dalo (the local name for taro), and mulberry (for tapa cloth). He also showed us some of the eight acres of vanilla beans that he said they'd pollinated by hand.

Pointing to a large plant next to me, he explained that villagers dried its long, narrow leaves and used them to weave their mats. He then found a small mulberry sapling, stripped off its bark and, skinning it, showing us the fiber used to make tapa cloth. He said it was first dried, then soaked overnight and beaten; the process could stretch an inch-wide strip out to a foot, or maybe even wider.

On our way back to the trail, he picked up a conveniently placed long pole and knocked down a green coconut. With his machete, he slashed off one end and then wiped the point of the blade on a piece of husk before poking a hole in the nut. As he wanted none, Dave and I shared the cool, sweet liquid inside. When Dave pointed to notches cut into the trunks of the coconut palms and queried, "For climbing?" Vaha nodded and smiled, saying, "But only for young men!"

We returned through the village and were back at the jetty by 1230. Vaha retrieved our fruit basket, which he'd placed in a locked shed, then asked if he could come and see our boat. Not wanting to risk swamping the dinghy, Dave made two trips out to *Windy Lady*. We showed him through the cabin, pointing out the radar, GPS and radios, and he then suggested we go outside, so he could have a cigarette; he had smoked continuously all morning. I wanted to offer him some refreshment but we didn't have much onboard, so made tea and spread peanut butter on homemade bread.

Dave didn't want to show up empty-handed for dinner, so asked if we could bring coffee. Vaha shook his head, saying, "The fruit and the meal are gifts from a friend, no charge, but if you truly want to help, you could make a donation to my church." He then explained, "Some members of my church are going to Neiafu for a meeting on Wednesday, and we could use the cash, but only if you want to help. It's really not necessary."

Dave nodded his understanding and soon politely suggested that it was time for the visit to end. Vaha thanked me for the food and drink, saying that he would go and start preparing for tomorrow.

That left us in a quandary, wondering just how much of a donation would be appropriate, but I had another problem, too, because I could not attend a Tongan church wearing shorts. I went through the cabin lockers, looking for anything that I could cut up to make a skirt, and decided that an old mauve and white, checked cotton sheet would have to do. I dug out my portable sewing machine and eventually, by trial and error, produced a garment. I don't think it looked like a bed sheet and it served the purpose, being roomy enough to wear over a pair of shorts, while covering my legs to the knees.

Earlier in the day, Dave had called Club Hunga on the VHF radio and made reservations for dinner. He then kept an eye on the mooring buoys out front, as he didn't want to be searching for *Windy Lady* on the far side of the lagoon in the dark. When he saw that one was free, we motored over and picked it up. We went ashore at 1800, and found fourteen other cruisers already seated at a long table, many of whom we'd already met. Dave and I were soon sipping on cokes at the happy hour price of $1 each.

Pete and Happi ran the restaurant. He was a fortyish Kiwi; she was Tongan and some years younger. The dinner was a set menu, priced at $12 each, but we found the food a little bland (fish, chicken and potato). Part of the meal, however, was cooked in an umu, an underground oven, and we traipsed outside to watch Pete start a bonfire and burn it down to coals, heating several large rocks in the process. He then placed the rocks into a shallow hole in the sand, dropped in small packets of food wrapped in leaves, and covered it all up. The packets were served with the meal, but whatever was inside remained a mystery.

After dinner, Happi visited with us for a while. She was an attractive woman and seemed to do everything at the club, cooked, sewed, cleaned, and ran the cruisers' net. She was well spoken and talked freely about herself, explaining that she'd left the village with her mother at age seven and returned only two years earlier. She added that it was proving difficult to adjust to village life, but at the same time seemed disdainful of the life style. She was particularly critical of the medical services, which she said were inadequate, and the churches, which she claimed had too much control over people's lives.

Next morning, we left *Windy Lady* on the mooring buoy and dinghied across the lagoon to the village. We thought we had plenty of time to make the 1000 service, but the prop hit a rock about fifty meters off the beach and broke the shear pin. Scrambling out in knee-deep water and wading through mud, we half-carried, half-dragged the dinghy ashore. Leaving it on a narrow strip of sand, we climbed over a low spot in the fence and hurried up the hill, but the sound of singing reached us before we made the top.

The singing grew louder as we approached the building that Vaha had pointed out the day before, and through an open door, I could see people standing inside. Cleaning the sand from our shoes as best we could, we removed them and stepped up over the doorsill. A man, standing in a back corner, edged over to let us squeeze in beside him. The singing continued for a few more minutes and then everyone sat down cross-legged on grass mats that covered the floor. We took the opportunity to move forward to a vacant mat, and everyone in the room watched as we sat down.

The minister started to read from the bible and, at the end of every verse, a deep rumbling sound came from the throats of the men; I wasn't sure whether it signified agreement or dismay. Meanwhile a couple of little girls, maybe four years old, moved back and sat and stared at me, not taking their eyes off me for at least half-an-hour. After the reading of the scriptures, we were on our feet again for more songs. Although they had no musical instrument to guide them, the congregation sang with gusto, never faltering. Men and women sang different parts, and the result was quite beautiful.

When we again sat down, the minister began his sermon. He was an older, grey-haired man and spoke quietly at first, but as he got into his message, his voice grew louder and more emphatic and his arms began to wave excitedly. I didn't understand a word but began to feel uneasy and started to look around the room.

The building was maybe twenty feet by forty, with corrugated metal covering the outside. Doors on both sides and the end were propped open with sticks, as were the rough shutters that covered the windows. I could feel the uneven surface of the pole floor beneath my ankles and noted that grass mats covered it wall-to-wall. A long, white cloth draped the desk up front; similar cloths covered two shelves located high on the wall behind it, each containing a kerosene lamp.

The congregation grew restless as the sermon dragged on, and now and then a woman slipped out with a child or a man disappeared. (I saw one through the open doorway, smoking a cigarette.) One woman had a woven grass fan with which she occasionally batted the heads or shoulders of two ten-year old boys, who obviously weren't paying attention. Small children walked back and forth among the seated adults, stopping briefly beside a familiar figure and then moving on. Several of the little girls wore shoes and socks inside and all wore colorful, lacy party dresses (most were pink.)

There were maybe twelve men in the room, about the

same number of women, and double that many children. The men wore black jackets over lava-lavas (wrap-around skirts) and most had on shirts and ties. Many of the women wore colorful shirts and long skirts, along with flat-brimmed, woven-grass hats held in place by long pins stuck through their hair buns. One woman stood out, though, as she wore an off-the-shoulder taffeta dress with long, dangling earrings and had on makeup. When she sat down, two lighter-skinned, blondish boys moved back to sit beside her.

When the preacher finished speaking, we rose for another song, and then a second man went forward and sat at the desk. Opening a ledger, he began calling out names, and with each name, a young child went forward and threw money onto the table. He counted it, called out the amount, and entered it into the ledger. The room was now very quiet, and I thought of Happi's criticism that the church had too much control over people's lives; certainly, such a public accounting placed a lot of pressure on them.

Singing from other churches drifted into the room, and I had the distinct impression that the rivalry amongst them included which choir could sing the loudest. The hour-long service then ended with another song. I slipped out with the first group but Dave lingered inside; he later said that all the men came over, shook his hand, and thanked him for coming.

The little girls attached themselves to me as soon as I was clear of the building, so I took some pictures while I waited. Everyone quickly disbursed, and as we walked away with Vaha, one of the little girls danced around us.

When I asked about her, he explained that he and his wife had adopted her when she was a month old. He added quietly that his wife had passed away two months earlier, so now his mother looked after her.

Removing our shoes outside the front door of the house, we entered a large, plainly furnished sitting room and sat down on a settee that occupied one side of the room. Vaha took possession of an easy chair at the far end and introduced his married daughter and 16-year old son, who sat on another settee directly across from us. After politely greeting us and asking where we were from, both quickly left the room. Meanwhile, our host had changed his clothes and the little girl, who appeared briefly, had changed out of her party dress.

The room was neat and tidy, and I noticed many small framed pictures sitting on a TV set in the corner and on a small table next to one wall. The lights were 12-volt fluorescent and were obviously powered by the solar panel; the TV didn't work. Through the open back door, I noticed an older woman hobbling back and forth and assumed she was Vaha's mother, but we never met her.

Another woman walked by the door carrying a plate of food, and shortly after the son entered the room. He laid a colorful floral cloth in the center of the large grass mat that covered the floor between the settees. Dishes of food were set on in, along with three plates, two forks and a spoon, and Vaha then invited us to eat.

Dave got down on his knees and filled one plate, asking about the food as he did so. The dishes included: tuna (cooked in an umu); corned beef and coconut cream, (wrapped in a leaf); breadfruit and tapioca root; papaya and coconut cream (hot). Dave handed me the plate he'd prepared and filled a second for himself; Vaha waited until we were eating before serving himself. The food was tasty but I found it very rich.

We talked about fishing, farming, and family life in Tonga, where Sunday was "for church and cooking only, to be spent with the family!" Vaha explained that his 16-year old son had four years of schooling and liked to spend time in Neiafu, with his girlfriend. I was asked if I'd like to take his picture and was happy to do so. With three studs decorating one ear, he looked like he'd fit right in at home.

We attempted to explain that modern life in North America was not without its problems, while emphasizing what we saw as valuable in their traditional lifestyle. However, I think the message was lost. They knew the hardships of their own lives and saw only that we were rich, with a big boat equipped with a diesel engine, radios, radar, lights and a stove.

About 1230, Dave thanked Vaha for his hospitality and gave him an envelope with a donation of $25 US for the church. Coming up with that amount had been difficult because we didn't understand the culture well enough to know what was appropriate. In the end, we decided to contribute an amount equal to what we'd paid for dinner at Club Hunga. With that, we left, finding our way down the hill to the dinghy, and Dave rowed back across the lagoon.

CHAPTER 13

Island Hopping

We left Hunga the following morning, motor sailing around the south end of the island and heading east. Two hours later, we dropped the hook in a quiet corner of anchorage #16, close to the island of Vaka'eitu. While giving *Windy Lady* time to settle, we made plans to snorkel on the ocean side of a reef that extended between two of the islands. We then loaded our gear into the inflatable and set off across the bay.

To start with, the water was choppy, tossing the dinghy about roughly as we landed, which probably should have warned us. But we just dragged the boat well up on shore and walked down to the end of the island. We then stood, staring in amazement; powerful waves were sweeping in across the reef, pounding the shoreline, and pouring through the gap. With strong winds and a high tide, we wouldn't be doing any snorkeling; in fact, with the sand stirred up, swimming in the lagoon wasn't even an option.

Next morning, the winds were still strong but the day warmed up nicely, so we returned to the gap at low tide. The reef was maybe 100 feet wide and the water knee-deep, so we carried our snorkeling gear part way across before stopping to put it on. Despite buffeting by wind waves, Dave managed without mishap and disappeared over the reef's outer edge. I wasn't so lucky, as a wave knocked me down while I struggled to pull on my second flipper.

The surging water then rolled me back and forth over the coral, and I couldn't get back onto my feet. When I banged into a chunk that was big enough to hang on to, I grabbed hold until the waves subsided. Only then was I able to pull off my flipper, get to my feet, and wade ashore. Shedding a few tears of relief/frustration, I discovered numerous coral cuts on my bottom and thighs. Deciding that I'd had enough surf for the day, I walked back to the lagoon.

I now pulled on my gear again, figuring that there had to be something worth looking at on this side of the reef. Visibility still wasn't that good, with fine sand suspended in the water, but I could see down about fifteen feet. Then, as the steep wall of the reef rose in front of me, I saw schools of small fish darting through the white branch coral that covered it. Upon closer inspection, the colorful streams were fascinating, ranging from brilliant blue to aquamarine, yellow, and even black and white.

For the next two days, the weather was gorgeous, exactly what you'd expect on a tropical island, and we even had the anchorage to ourselves for twenty-four hours. With warm sunshine and 10-kt breeze, we spent our time on the beach and in the sea, searching for shells and snorkeling.

Dave again swam on the ocean side of the reef, telling me that the outer wall plunged down a long way, and he'd seen several very large fish. I was content to stay in the lagoon and learned that on a rising tide, ocean waves first pushed in very warm water sitting on top the reef. Cooler seawater followed soon after, bringing refreshing currents that were pleasant to swim in and drew all manner of small fish into the humps and fissures on the face of the reef.

The following day, we visited an eco-resort situated on a hill across the bay. Their dock was just a low raft, anchored well out from shore, and with wind waves breaking on the side, we were again dodging salt spray. After securing the dinghy, we climbed up onto a rickety catwalk that took us across 200 feet of shallow water. We then stopped to inspect a dugout canoe sitting on the beach before following a broad path through the trees and up the hill.

The resort occupied an idyllic spot with scenic views of both the lagoon and the ocean channel on the far side. All the buildings were made with native materials, so had grass-mat walls and thatched roofs. As we wandered through the grounds, we checked out several small sleeping huts (fales), the restaurant, a bakery, and an outside bar. The only person we saw was the bartender, an Austrian named Hans. After talking to him for a while, Dave asked if we could get a meal. Hans shook his head, saying,

"You need to book and it's too late for today." He then disappeared momentarily and on his return announced, "I've talked to the cook and she can have a meal ready at 1830."

We returned to the boat for a few hours and climbed the hill again at sunset. As we were early, I sat on a swing outside the bakery, watching the light leave the western sky while darkness crept across the ocean channel below. A colony of fruit bats flew through the trees, disturbing the evening quiet, and a few minutes later, a young boy appeared and called us to dinner.

He led us into a room lit by kerosene lamps and furnished with wooden tables and benches. Austrian music played for a while, but soon the room fell silent; we were the only customers. The meal was delicious and included a salad of lettuce, tomatoes, and carrots with a tasty dressing, and a main course of tapioca root and mahi-mahi in garlic sauce. We paid $10 each.

After dinner, we visited with Hans and his Tongan wife, Melee, who was the cook. Knowing the country had a king, we asked about the political situation and Melee responded, "We have a good King and people are generally satisfied with the government." Hans then explained, "Every male child receives eight acres of land upon reaching maturity, so as long as land is available, people will probably be satisfied." We left at 2030, having very much enjoyed the food and conversation, and made our way down the hill with the aid of a flashlight.

Overnight, the winds backed to the north, and by morning, the sky was hazy, the water cloudy, and bits of debris floated around the boat. The haze increased as the day progressed, and by late afternoon, the sun was an angry red disk on the horizon. We returned to Neiafu about midday, and so set a pattern for our stay in the islands, as we preferred to be in the village whenever the weather turned stormy.

As soon as the weather cleared, however, we'd be off, usually traveling at midday when visibility was at its best. Dave would take the helm, following winding routes between the islands, while I stood watch at the bow, looking for changes in water color and searching for buoys marking the passes. As we explored different parts of the archipelago, we anchored in turquoise waters near palm-clad islands with white sand beaches. If we happened to be near coral, we took extra care not to cause damage.

We didn't have a favorite anchorage, but preferred those that were sheltered and not too busy, so returned to Lisa Beach (#10) on numerous occasions. It had only a small beach but was just half-a-mile away from a snorkeling site off Mala Island. When we awoke to a clear, bright sky one morning, with a light breeze and low waves in the channel, we loaded up the inflatable and headed across.

The next two hours were extraordinary and proved to be the highlight of our visit to Tonga. With a low tide, the warm, crystal-clear waters in the passage next to the island were five to six feet deep. The white sands on the bottom were dotted with colorful clumps of various types of coral, including coral heads. With the sun high in the sky, the views in this beautiful underwater garden were like nothing I'd ever seen before. I was enchanted!

The corals came in every color and shape imaginable, including yellow, blue, white and pink; there was blue-tipped branch coral, white branch coral, yellow fan coral, brain coral and table coral. The fish were just as fascinating, appearing in solid shades or with stripes, bands or dots, and their colors ranged from black and white to pale yellow, deep orange, brown, aquamarine, brilliant blue and silvery white. It was here that Dave picked up a large conch shell and then later, at Lisa Beach, he found a smaller tiger cowrie shell.

We returned to Lisa Beach a few weeks later, after spending several stormy days in Neiafu. We checked out a couple of anchorages on the way by, but they didn't seem as protected and we thought the weather still unsettled. Three boats were sitting peacefully at anchor in one of them, including a beautiful, black sloop named Scaramouch that we'd previously seen in the village. It was a forty-eight foot Swan, of Danish registry, and was apparently under lease to a group of five men, all of whom were sitting on deck, playing cards and drinking beer.

We spent that afternoon and evening quietly on our own, but around 2200, well after dark, the roll of thunder reverberated through the cabin. Going up to the cockpit, we watched spellbound for half an hour as a storm moved in. Sheet lightning flaring behind the cliffs on one side of the cove provided a magnificent light show to accompany the deep drum rolls of thunder. As the storm advanced, the occasional flash of fork lightning split the darkness on the other side with an intensity that was almost blinding.

We only went inside when the rain started to bucket down, and while we didn't notice the wind, *Windy Lady* then turned and turned on her anchor chain. At daybreak, she was barely fifty feet from a cliff face but still in sixteen feet of water; the dinghy held a foot of rain.

As usual, we tuned in the cruisers' net after breakfast, and that morning the news practically jumped from the radio. We could only stare at each other incredulously as a disembodied voice reported, "You probably know that Scaramouch was lost last night. The boat went up on a reef during the storm, rolled onto its starboard side and was holed. The crew is safe, and villagers and other cruisers worked with them throughout the night to salvage what they could." As the words echoed in the stillness of the cabin, we couldn't image what had happened. We'd seen the boat in safe harbor only hours earlier.

Going up to the cockpit, I stood and looked around the bay, seeing a gorgeous morning with bright sunshine and sparkling water. What a difference a few hours could make! In my mind's eye, I tried to picture the events that had taken place while we'd sat outside watching the storm: wind and waves battering the Swan, the coral reef grinding on the hull, a desperate crew.

Eventually we learned that the crew had planned a trip to the Ha'apai Islands, which were a full day's sail away. Wanting an early start, they crossed through the barrier reef near the southern island of Ovaka and anchored on the outside, thus exposing themselves to the full fury of the storm. A high price had been paid as a consequence of their actions, but there seemed to be almost an inevitability to the tragedy, as if events had taken on a life of their own.

We were pretty cautious in reef country, probably because of the scare we'd had in Palmyra, but after we'd gained some experience, we planned a trip to Kenutu, anchorage #30, on the eastern edge of the archipelago. It was a half-day journey from Neiafu, and we set off one bright, sunny morning. The route wound between islands and across reef-dotted shallows, and before long, I was aware that the occasional channel marker was missing. Knowing that walls of coral rose up almost vertically from the bottom, my stomach tended to knot up whenever the depth gauge started to rise.

Then, as we crossed an underwater ridge near the end of our journey, the water slowly got shallower and shallower. At fifteen feet, my eyes were glued to the depth gauge; at ten feet, I started holding my breath; at eight feet, with only two feet of water beneath the keel, the bottom finally leveled off. A few minutes later, we dropped the hook in the sheltered waters of the lagoon.

Kenutu was a barrier island and the scenery was spectacular. Huge Pacific Ocean swells crashed onto reefs running between it and two adjacent islands, while nearby, ten sailboats floated serenely in the quiet lagoon. Mesmerized by the sheer power on display, we walked the beach for hours, and at low tide, strolled across the scoured surface of the reef itself.

When the ocean poured back into the lagoon at

high tide, we took to a trail that wound up the hillside, walking in soft, greenish light as sunlight filtered through the canopy. The vegetation changed at the top (about 300 feet in elevation), with evergreen trees covering the windward side of the island and the ground carpeted with needles. Looking out over open water, we glimpsed a wide coral reef extending off to the southeast, but my eyes kept returning to the spray that flew high into the sky as waves pounded the rocks below.

On our way back to Neiafu, we overnighted at Port Morelle. Next morning, we awoke to sunny skies, light winds, and calm seas, perfect conditions in which to visit a local attraction known as Swallows' Cave. So, we dinghied down the channel, where swells swept us in through a narrow fissure in the cliff face.

We found ourselves in a large cavern, and I first noticed the rich colors, the oranges, greens, purples, and golds that coated the walls above and below the water's surface. I then studied the high domed ceiling and hanging rock sculptures edging its perimeter. We found smaller, semi-circular rooms at either end of the main chamber, and reportedly, another large room was behind them. With a floor above sea level and a hole in the ground that admitted light, it had been used for ceremonial purposes in times past.

What most fascinated me though was the quality of the light in the cavern. Coming from below, it was filtered through water that was crystal-clear and incredibly blue. Peering down, I wondered how deep it was, and then something far below caught my eye. Looking closer, I saw a flash and made out the fluorescent markings on the fins and wetsuits of four divers. Other than the markings, I could see only their outlines and figured they were down about forty feet.

The divers were off a trimaran anchored close by, and as they passed, a trail of small bubbles rose to the surface. I then saw a giant bubble come off the sea floor; it rose steadily to the surface, where it formed a boil about two feet in diameter. I had no idea what caused it, but several more followed.

In early October, we anchored near a white sandy beach on the edge of a broad expanse of turquoise water. We did some snorkeling in a nearby channel, where the coral heads were three to four feet high and spread out like tables in a dining hall. But the tide was too high, so we planned to go back next day at low tide. We then contented ourselves with strolling down the sandy beach, splashing in warm waters, and admiring the views. When the winds came up overnight, stirring up the sand, we spent the next day doing chores.

The following day, the world caught up with us, and it wasn't even the modern world. The tranquility of our morning was shattered at 0845 by a sudden loud rapping on the hull; quickly repeated, the knocks conveyed a sense of urgency. Dave rushed out to the cockpit, where he saw a woman sitting in a dinghy; she was off *Southern Cross*, one of the boats anchored nearby. Handing up a piece of paper, she said that she'd picked up a message for us on a HAM network. We would later learn that the message originated with Peter on the Triple D net in Victoria and was picked up and relayed by *Cabezon*, *Kiley,* and *Hope II.*

The note read, "Phone home immediately, Mrs. Janitsky very ill."

Knowing that more bad news awaited, we got the boat organized and underway; when we anchored in Neiafu several hours later, winds were gusting at 20 kt. After a wet dinghy ride ashore, we walked down to the Communication Center, where I stood in line; my call was put through at 1430 local time (1830 of the day before in BC).

The conversation that followed was difficult, made more so by a delay in voice transmissions that had us talking over each other's words. However, the message was clear. If I wanted to see my mom and say goodbye, I needed to go home immediately. I knew that wasn't possible and had reconciled myself to that fact before leaving Canada. Unhappily, I now had to convey that message to the folks back home.

Gusty winds howled through the rigging that night, causing *Windy Lady* to pull and twist on her anchor chain. Rumbling noises echoed underwater as the chain dragged across the bottom, but the anchor held. Next day, 25-kt winds pushed up high waves, so we didn't even go ashore. Wind and rain persisted for a few more days, so we stayed in Neiafu and I called home several times.

When we grew tired of sitting in the village, we moved to a small, sheltered cove not far away. A charter boat was anchored nearby, and just before dark, Dave noticed a dinghy drifting silently away from the stern. Giving chase in our own dinghy, we returned it before the crew even noticed it was missing, so were able to pass on the good deed done for us by Bob on *Adios* in Hilo.

We visited with the crew for a while, learning that they were part of a group of twenty-nine Brits, who had chartered five boats for periods of two to four weeks. They had made their reservations eighteen months in advance, and some were not very happy with the cool, rainy weather. I sympathized, thinking I probably wouldn't have been very happy either, if the previous week had made up half my vacation. Fortunately, we weren't on holiday; whatever happened was just part of the journey.

Actually, aside from the occasional storm system tracking through, we found the weather in Vava'u very pleasant. Afternoon temperatures were now in the mid 80's, although I was pulling up a blanket during the night. What we found tedious, however, was the regular use of the word "fine" in weather forecasts because the sun often disappeared after a few hours, leaving the day cloudy, windy, and rainy.

As cruising season drew towards an end, boats started to leave Vava'u, and one of them was *Byjingo*. A few days before they left for Australia, we joined Dave and Maree for dinner on their boat. We were in Neiafu at the time but anchored half a mile apart, so dinghied down just before sunset. Heavy, dark clouds moved in while we were there, and the night was pitch-black when we said goodbye a few hours later.

I have no idea how Dave found *Windy Lady* that night because we couldn't see a thing. The boats at anchor didn't show up until we were almost on top of them, then one would suddenly loom out of the shadows and he would steer around it. Peering ahead, I looked for anchor lights, but they seemed disconnected and floated in the darkness like dim stars. A few boats had cockpit lights, which made them easy to spot, but we didn't have one and I wondered how we'd find *Windy Lady*. He then turned the bow twenty degrees and there she was.

As we started making our own plans to move on, I thought back over the weeks we'd spent on the hook. We'd done most of our living in the cockpit, even showering there after dark, as it was much less humid than in the head. The house battery bank had provided a bit of power for lights, but normally, it had been early to bed and early to rise, with not even a barking dog breaking the silence.

The dinghy had become an integral part of our lives, and Dave regularly scrubbed the bottom and serviced the engine. It also needed frequent bailing and, early on, we scooped out three buckets of rainwater, which gave us our initial supply of wash water. After that, we collected enough rainwater from the tarp over the cockpit to meet those needs, which meant that the water in our tanks was used only for cooking and drinking.

We became adept at washing clothes with bucket and plunger, hanging sheets and towels on a line strung across the foredeck and pegging other items to lifelines. Usually, even heavy shorts and towels dried quickly. Afterwards, I used the rinse water to scrub the cockpit and always puzzled over how so much dirt could track onboard, especially when the boat was sitting in the middle of a lagoon.

The wet market had been one of the reasons for our regular visits to Neiafu and we'd bought fresh tomatoes, cucumbers, green peppers, papayas, bananas and coconuts. Large papayas cost 50¢ each and a three-pound bag of sweet, juicy tomatoes was $1.00. While there, we disposed of a bit of garbage and sometimes bought potatoes and onions in the shops.

We usually anchored near the landing used by the villagers and, early on Friday mornings, had seen a few small, overloaded boats come chugging into the lagoon. As Friday's were market day, villagers had been crammed into every available space on the boats, including on top the cabin roof. Later, produce would be piled high in the stalls at the wet market, with the plaza outside filled with stacks of large roots, leaves and melons, as well as watermelon, cabbage, spinach, lettuce and even a few potatoes.

We had wondered about the health of fish stocks when we first arrived, and decided that, whatever the numbers, we would leave them for the villagers. So when I saw four men hauling in a net early one morning, I watched curiously. They seemed to remove many small fish, but I only noticed two or three that were big enough to fill a frying plan.

We then heard about ciguatera, a debilitating illness that came from eating fish that fed on coral, at which point our decision seemed even more sensible. We met two cruisers who'd been seriously ill with it, and both men explained that the effects were cumulative, so a second attack could be life threatening. Apparently, the bigger the fish, the more toxins it accumulated, but villagers seemed to have developed an immunity.

I thought the dark-haired, brown-skinned Tongans a handsome people. Not wanting to be rude, I tried to study them unobtrusively and never took pictures without permission. The men all wore shirts and long pants, with soft, short, woven-grass mats wrapped around the hips and tied at their waists. The women, both young and old, wore long dark skirts, usually with tunic tops, but on market days, a few younger women stood out like butterflies in bright, colorful muumuus.

Schoolchildren were always noticeable as girls and boys alike wore uniforms with white tunic tops and blue wrap-around skirts. And, on one occasion, I saw a woman with a long grass mat tied around her middle. The mat extended from waist to ankles, so stuck out straight behind when she bent forward.

However, I had felt uncomfortable in the village, as I seemed to be under constant scrutiny. Assuming the villagers didn't approve of my clothes, I started wearing my baggiest shorts and loosest t-shirts ashore. When that didn't make a difference, I tried to ignore my discomfort, but it never completely away. In fact, one morning, when the town was busy, with many men sitting on benches in front of the storefronts, I'd complained that I felt like I was walking a gauntlet.

It never occurred to me that my coloring might be responsible; that my short, blondish-white hair and blue eyes were just unusual to their eyes. But years later, in Asian countries, total strangers asked me to pose in their pictures. The first time it happened, I thought the woman was asking me to take a group photo, but no, she wanted me in the group. I then realized that I did something similar myself. I stared at people with very black skin, fascinated, because they seemed to glow from within.

CHAPTER 14

Planning Angst

Most cruising boats leave the tropics before the start of the South Pacific cyclone season, which runs from November through April. So, as October rolled on, yachtees planning to make the trip to New Zealand met and compared notes. That was when many of us first heard of the Queen's Birthday Storm. The memory was recent enough to have left a deep impression on those who knew about it, but for others, like us, the story was a bombshell.

Just two years earlier, at the end of May 1994, some thirty-five boats left New Zealand, heading for Tonga. Reports indicate that on June 2, sailing conditions were almost perfect, with trade winds well established. On that same day, however, a small area of low pressure between Fiji and Vanuatu began to deepen and move southward. It moved rapidly, expanding quickly, and by June 4, many of the boats in an area south of Minerva Reefs were directly in its path.

The system brought gale and storm force winds over

a 900-nm radius, so was impossible to avoid. Winds of 50–60 kt and high seas battered many of the boats. When it was over, one yacht had been lost, along with three crewmembers, and seven boats were abandoned. Rescue efforts, coordinated by New Zealand, involved ships of four different nations.

The impact of the story was palpable, and cruisers turned to their friends to talk it through. Unfortunately, the more they talked, the greater grew their anxiety. It became a vicious circle, where everyone fed off everyone else, and a couple of doomsayers in the group kept the pot boiling. The rhetoric became so bad that some yachtees talked about spending cyclone season in Fiji, or hiring someone to sail their boat to New Zealand, or maybe even selling their boat.

As if that wasn't enough to deal with, other stories circulated that played on our fears and further eroded confidence. One such story concerned a cruising couple like ourselves, who ran into trouble a few days out of New Zealand. The man apparently fell overboard during the night. When his wife went looking for him in the morning, she found his tether, still attached to the harness he'd been wearing, dragging the body behind the boat. She couldn't get him back onboard, but managed to carry on until close enough to call for help.

I told myself that these stories didn't change anything. We would prepare as best we could and then go to sea and deal with whatever happened, when it happened. But when the old specter of a storm at sea returned to haunt me, I knew my confidence was being undermined. Dave was impervious at first, saying only that the dithering was painful to listen to, but finally it got under his skin too. In exasperation, he groaned, "You'd think no one had ever made the crossing to New Zealand before!"

Other than making a rule not to travel in cyclone season, we hadn't worried much about weather up to that point. Anything could happen during two weeks at sea, so we simply picked a day, maybe waited for stormy weather to clear, and then left. In fact, during our crossing of the Pacific, we had gone without weather forecasts for weeks at a time.

While we were familiar with cold fronts, warm fronts and such, we didn't know anything about weather patterns in the South Pacific, so when a few cruisers organized a weather seminar, I persuaded Dave to attend. Unfortunately, we didn't learn anything, as the group just sat around and talked about weather windows and strategies. However, I did purchase a copy of a weather manual prepared by meteorologists in New Zealand; it proved very informative and I studied it from cover to cover.

We also met two couples who had made numerous crossings between New Zealand and Tonga. One couple was on a large ketch named the *Maggie Drum*; they'd made the crossing seven times. The other couple was on a boat called *Sanctuary,* and they'd crossed five times. Neither could explain the current furor, saying only that their passages had always been very normal.

Still, I needed to do something, so started monitoring HF weather broadcasts from Hawaii and New Zealand.

Once I was familiar with the format and information provided, I began copying down the details and rediscovered the shorthand that I hadn't used in twenty-five years. Knowing that we could receive weather information at sea boosted my confidence, but there was still one difficulty, as radio transmissions could be garbled over long distances.

Dave finally had heard enough and decided to leave for Nukualofa, Tonga's capital city, which was an overnight trip to the south. It was also the jumping off point for the passage to New Zealand, so we got busy and finished our maintenance work. But, a couple of days before our scheduled departure, we returned from a trip ashore and heard the bilge pump running.

When he investigated, Dave discovered that a fitting on the salt-water strainer, which was part of the engine cooling system, was leaking; he needed a three-inch long piece of threaded ¾-inch copper pipe to repair it. We didn't have anything suitable on the boat and didn't know where to start looking ashore, as there wasn't a marine store or even a hardware store on the island.

We checked at a plumbing supply store, with no luck, and talked to several different people, all of whom gave us little comfort. Finally, one fellow suggested we go see Carter Johnson, the owner of the Paradise International Hotel. After traipsing halfway down the lagoon, we found Carter tending his garden at the back of the hotel. Actually, he was standing over some severely damaged tomato plants, berating himself for having confused imperial and metric measurements when mixing up some bug spray.

He politely listened to our tale of woe and then led us over to a small shed at the back of the property. Directing Dave toward a bin full of galvanized fittings, he told him to have a look. Within minutes, Dave found exactly what we needed, for which he paid one pa'anga, about $1 US. Feeling elated, we returned to the boat and Dave replaced the fitting, while I thought about how uncomfortable the passage might have been, if the fitting had failed at sea.

This was the third time that equipment had failed just as we were heading out. The pad eye broke on the boom near Cape Flattery, the winch flew apart in Palmyra and now a fitting had corroded through. Having pondered over *Scaramouch's* fate, I was starting to think that maybe the captain's luck transferred to the boat. I had always considered Dave lucky and wondered whether that might be something to consider, if you were thinking about going to sea.

We needed a clearance document to sail between Vava'u and Tonga's southern islands, so cleared with Customs and Immigration on Friday, October 18. We intended to leave the following morning but a trough moved in overnight, bringing southerly winds of 30 knots and rough seas. Having no desire to fight headwinds, we waited it out. For the next week, the weather was terrible: cloudy, rainy, windy, and cold. When the winds eventually shifted to the east, still at 25 kt, we moved the boat to Hunga Lagoon, where we could get an early start the following morning.

After picking up a mooring buoy in front of Club Hunga, we spent a few hours putting out jack lines and safety harnesses, and ensuring that everything was properly stowed and tied down. We then visited the restaurant and celebrated my fiftieth birthday, which had fallen the day before; it still seemed appropriate, however, given the twenty-hour time difference.

Having a few beers the night before going to sea wasn't really a good idea, as I was conscious of a dull ache behind my temples when Dave rousted me out at 0530. We then ate breakfast in the muted lights of the 12-volt system before preparing the galley for sea. The sky was still dark when we untied the mooring line and motored across the lagoon. In the greyness of early dawn, we were out on deck, raising the mainsail, and as soon as it was light enough to see, we transited through the reef.

Just after 0700, on Saturday, October 26, we turned the bow to a heading of 195 degrees. With east winds steady at 20–25 kt, we had two reefs in the mainsail and one in the jib. For the first time, we also raised the staysail, which increased boat speed by half a knot. Seas were very rough, with spray flying into the cockpit as eight-foot swells broke along the port side. Before long, 30-kt gusts were bringing even larger swells, causing the boat to heel sharply as they swept beneath the keel. With all that sail up and lumpy seas, Otto needed constant attention, but we figured we could handle it for twenty-four hours.

Within an hour of leaving Hunga, a small pod of humpback whales surfaced about 100 yards away. It was a rare treat to see these fascinating creatures in their natural environment, and spellbound, I watched until they disappeared from view. We were aware, however, that incidents between whales and yachts were not unprecedented. *Windy Lady* actually seemed a little vulnerable when close to such large and powerful animals, as most adults were at least as long as she was and more than double her weight.

By mid-afternoon, we were in the lee of the Ha'apai Islands and seas were noticeably quieter. Although we had talked about stopping, we decided to carry on; we were enjoying the passage and had no qualms about sailing at night. During the watch ending at 2000, we logged twenty-six nm, and by then, had left sheltered waters behind.

Winds veered to the ESE shortly after dark, settling at 15–20 kt, and seas were then about five feet. Soon after, a full moon climbed above the clouds on the horizon, and the hours passed quickly as *Windy Lady* raced through the night. At 0730, we reached the GPS waypoint marking the start of the channel into Nukualofa harbor; we'd made good 150 nm in twenty-four hours, something of a record for us.

We now stared in confusion at the open water around us; not even a buoy marked the spot, although a couple of small islands were visible in the distance. Dropping the sails and starting the engine, we spent the next three hours plotting positions, following headings and studying the color of the water. On three different occasions, the depth gauge flashed a reading of eight feet, when we were actually in one hundred feet of water. Dave was on the helm and hadn't slept all night, so reacted each time by shoving the gearshift lever into reverse.

After motoring for fourteen nm, we safely arrived at the small boat anchorage off the harbor, where boats were stern-tied to a seawall. With 20-kt winds and only a few empty slots, Dave faced a difficult job of backing the boat into position. However, as soon as we dropped the hook at the edge of the channel, three dinghies appeared from nearby boats and pushed the hull around for us. Before *Windy Lady* had even settled in place, the young man who had taken our mooring line was knocking on the hull and asking if we wanted crew for the passage to New Zealand.

We didn't, had never even considered it, but now learned that a floating population of young people, men and women alike, traveled around, working as crew on sailing boats. As positions or people became available, notices were posted on bulletin boards outside ablution blocks at major departure points. In fact, the boat next to us, *Sojourner*, had taken on two crew members, so Pete and Camille were able to tell us something about the chap who'd come knocking on our hull.

A few days earlier, two young men had approached them, looking for positions. Both were then working on another yacht but were desperate to jump ship, saying they thought the skipper unqualified and didn't trust his judgment. *Sojourner* had already taken on a crewmember, so had room for only one; the other man was the one who talked to us. The tale became more complicated in the days that followed with reports of unpaid wages and missing passports; eventually the local police had to sort it out. (Apparently, it was normal for a captain to hold onto the crew's passports, as he was responsible for everyone arriving on his boat.)

A rainfall record was set on our first morning in port as eight inches of rain fell in three hours. With 30–40 kt winds reported outside the harbor and visibility down to a quarter mile, I was just thankful that we'd arrived the day before. When the rain ended, we walked downtown, splurged on greasy hamburgers, then checked out the market and located our mail. After another day of rain, the weather improved, and we spent our days exploring the city.

We never took the same route twice and were amazed by the number of children we saw, along with the numerous schools they attended. We even caught a glimpse of the Tongan king on the main street, as he passed by in a small procession made up of two motorcycles and three cars. We shared that experience with Leo, an American single-hander, who was thrilled at this glimpse of royalty.

At the National Cultural Center, we learned that Tonga was the only country in the South Pacific that had never been under foreign control. The first Europeans to make contact were the Dutch, followed by the English. Captain Cook named them the Friendly Islands when he arrived in 1776.

The mutiny on the *Bounty* took place just off the Ha'apai Group in 1789, and it was from there that Captain Bligh started on his 3,618 nm voyage to Timor. Knowing that they had sailed in an open boat, and he had used only a sextant and pocket watch to navigate, I again marveled at the skill and courage of those early mariners. We had GPS, compass, depth gauge, detailed charts, and a diesel engine to keep us safe.

We came to know several of our fellow cruisers while in Nukualofa, one of them being the American, Leo, who was 74 years old. He was a small man, slightly balding, and we were most impressed with his vitality. Long after most folks had thrown in the towel, he was out there, still sailing the oceans. Within minutes of meeting him, he'd told us his whole life story; he'd circumnavigated the globe two and one-half times, lost a yacht on a reef in New Guinea, and had a prostate operation. In our numerous encounters, Leo never stopped talking, not even to take a breath, so it was impossible to get a word in edgewise. He would apologize, saying he couldn't help it; he was just making up for all those days and nights alone at sea.

We also met a couple from New Zealand who were our age and had been cruising for seven years. When five days out of Nukualofa, Graham and Jillian lost the self-steering on their boat, so took turns hand steering until they were safely in port. What made that so remarkable was that Graham suffered from detached retinas and was virtually blind. He wasn't able to see the bow of the boat from the cockpit but could see the compass well enough to hold a heading. When Dave asked Jillian why they hadn't sought medical treatment before starting the passage, she shrugged her shoulders and said, "We were in India when the first symptoms appeared and didn't trust the doctors."

Instead, they chose to make the long passage home before seeking treatment.

I, on the other hand, couldn't bring myself to phone home. I had last talked to my dad the day before we left Neiafu. Then, in the early morning at sea, I'd sensed my mom's fight was over. After mentioning it to Dave, I buried the knowledge, not able to face it. Although we located the Communications Center in a small building tucked into a backstreet, several days would pass before I could bring myself to place the call. Prepared though I was, I broke down in sobs when my sister, Pat, confirmed the news. Later, I stood on the street in front of the building, weeping into Dave's shoulder; the past had overtaken the present, and I felt like I had lost my entire family.

CHAPTER 15

Passage to New Zealand

With the coming of November, a few sailboats started slipping out of the harbor, heading south one by one. We weren't in any rush, as Dave figured it was still cold in New Zealand. But as more boats arrived from the north, the endless discourse about weather windows again intensified. Everyone was then so busy consulting with everyone else that many weren't able to make the decision to leave, a situation Dave aptly described as "paralysis by analysis".

We cleared with Customs and Immigration on Thursday, November 7, intending to leave the next day. Only then did we realize that we'd be starting our voyage on a Friday, which was bad luck according to an old maritime adage. While telling ourselves that we weren't superstitious, we certainly didn't want to jinx our good luck, then decided that we had no good reason to ignore maritime custom and stayed over a day.

We were hoping to make the 1,083 nm crossing in ten days, but once again, our route would not take us in a direct line. We would first sail south for fifty nm, which would have us clear of all reefs and other hazards. We would then turn the bow southwest, toward a waypoint located north of New Zealand. From there, we would steer south-southeast to Opua on the North Island. This approach would give us a better angle at the end of the voyage, as we wouldn't be pushing into any storm system that might be sweeping eastward across the Tasman Sea from Australia.

At 1000 on Saturday, November 9, we motored out of Nukualofa harbor, one of four boats to depart that morning. Two hours later, we were clear of Egeria Channel and stopped to raise the sails; winds were then at 15 kt from the NNW, with an overcast sky and light rain. Conditions quickly deteriorated, and when the island of Tongatapu disappeared from view at 1500, winds were howling out of the north at 20–30 kt and swells were about six feet. With winds and waves behind us, *Windy Lady* was making almost seven kt and becoming increasingly difficult to control, so we stopped and dropped the mainsail, then left out only enough headsail to provide steerage.

At 1600, Dave went below to rest and I stood at the helm, deafened by the roar of the sea, and watched spray and foam from breaking waves sweep past on either side. The sky now darkened as clouds grew heavier, and the seas continued to build. Before long, twelve-foot-high swells were thundering down around the boat, and then 35-kt gusts began pushing up even bigger waves.

The first time *Windy Lady* slewed sideways as she surfed down the face of a wave, I wasn't too concerned; the bow was easy to bring back around. But when it happened a second time, I belatedly recognized that we could be dangerously close to broaching. Horrified, I glanced at the grey, turbulent waters racing past and desperately tried to remember what I'd read about steering a boat in following seas.

I knew that I shouldn't steer straight down a wave, so tried to hold the bow at a slight angle. The problem was that once the bow started to turn sideways, I couldn't stop it. Somehow, I had to anticipate that movement and correct it before it happened. All went well for a while but then, in the fading light, I heard a loud roar coming up from behind. This wave was accompanied by a wind of 37 kt.

Once again, despite my best efforts, *Windy Lady* slewed sideways. Only now, she heeled over so far that mast and sail were lying out over the trough, and I was looking down into the abyss. The next moment, the wave swept underneath, rolling the boat back in the opposite direction—and my heart started beating again.

Beyond terror, I clutched the helm ever tighter and focused so intently on steering the boat that I blocked out much of the noise around me. I heard only the wave coming up behind and saw only the one directly in front. I soon realized that if I hoped to anticipate movement in the bow, I needed to be able to feel any change in pressure on the rudder, so loosened my death grip on the wheel.

I stayed at the helm for two more hours, holding *Windy Lady* in a delicate balance; sometimes, in a weird way, even relishing the challenge. Darkness came early because of the heavy cloud cover, and as the end of my watch drew closer, I started to think that maybe the worst was over. But minutes later, another loud roar came out of the darkness behind me; it grew and grew until it sounded like a freight train coming down on top of us.

Not daring to look around, I tightened my grip on the wheel, knowing we were about to get pooped. I don't know how I knew but seconds later, a wall of water arced over my head and fell into the cockpit and companionway. I emerged unscathed, wet only from spray, while a foot of water surged around the cockpit sole, then escaped out through the scuppers. Fortunately, Dave had closed the top of the companionway hatch when he went off watch, so not a lot of seawater went below.

At change of watch at 2000, north winds were still gusting at 20-30 kt, and seas were about ten feet. As we were then clear of all hazards, Dave turned the bow fifty degrees and we headed towards New Zealand. Just before midnight, the breeze backed to the NNW, putting us on a beam reach, and waves now broke alongside the hull, occasionally sending a shower of spray into the cockpit. Not a lot of water was coming aboard, maybe a glassful at a time, but for the next four hours, every splash hit me directly in the face. Just after daybreak, a front passed overhead, bringing rain and a wind shift of 120 degrees;

then the wind died. We started the engine at 0900, and at noon, after twenty-four hours at sea (Day 1), logged 107 nm.

After plotting our position on the chart, I checked the numbers for my wild ride the night before; I'd sailed twenty-six nm. It only then occurred to me that, given our boat speed, the 37-kt gust that hit us would actually have been closer to 45. I was just thankful for the 8,000 pounds of lead in the bottom of *Windy Lady's* keel that kept her upright and safe.

A 6-kt breeze rose out of the SSW that afternoon, teasing us into raising the sails. It died three hours later and we motored for the next forty-eight hours. The days were sunny and warm, the winds light and variable, and the sea grew calm, almost flat. With a dark moon, brilliant stars filled the night sky, some reflecting on the quiet water around us. Even with the noise of the engine, the calm conditions made it easy to adjust to napping three times a day.

We spotted a pod of humpback whales just before noon on the third day, this time maybe a dozen animals. Although they drew closer than the ones we'd last seen, they still didn't cross my comfort zone. Watching with admiration and wonder, I kept thinking, "So far from land, but so much at home!" For Days 2 and 3, we logged 87 and 103 nm.

We started sailing again shortly after midday, when a light breeze rose out of the NNW and settled at 8 kt. With winds almost on the beam and a three-foot swell rolling silently beneath the hull, *Windy Lady* cut smoothly through the water for the next ten hours. Otto was working well, so we spent the afternoon basking in warm sunshine, with the silence broken only occasionally by the sound of a wave curling off the bow.

At the start of my watch at 1600, I stood and slowly turned around, studying the perfect, unending horizon, where gently rolling blue waters met a pale blue sky. Later, I sat and watched the golden globe of the sun drop down to the edge of the world and then disappear. When I searched the sky, looking for stars, I noticed that the breeze felt cool on my face, much cooler than it had been in a long while.

That was when *Night Music* issued its first advisory, warning all boats to stay north of Latitude 25S. *Night Music* was the network used by the boats that had sailed from the US east coast. We had registered with them because I was worried about not getting weather broadcasts at sea. As a result, we now responded to a roll call once a day and listened to weather briefings that originated from a guru in Ontario.

However, we were only forty nm from Latitude 25S and, given the weather we'd been experiencing, found it difficult to take the advice seriously. I did go back to my notes from the morning forecast and sketched out a rough drawing, which showed gale force winds much farther south. Dave was more concerned about the fuel we were using, and shook his head in bewilderment, saying, "We're still a long way from New Zealand; we can't sit out here motoring!"

As darkness settled across the ocean, the wind started

to back and then dropped away, leaving the sails hanging limply and the boat rocking gently in the swells. Dave started the engine at 2200, and two hours later, the winds picked up from the WSW, right on the nose at 8–15 kt. The accompanying storm darkened the sky, bringing driving rain, and then the autopilot quit working. So I stood out in the rain, peering into a pitch-black night, and hand-steered. Winds and seas then pushed the boat straight south, and just before 0400, we crossed Latitude 25S. An hour later, the breeze dropped below 10 kt, the rain stopped, and Dave got the autopilot working again.

Next morning, *Night Music* again issued a warning for boats not to go farther south until westerly winds had cleared. Dave again shrugged off the advice, saying, "If we don't go south, we'll never get to New Zealand!"

At 1028, we crossed the International Dateline (180 degrees of longitude), and soon after, the winds increased to 10–12 kt. After sixty hours of motoring, we were tired of the sound of the engine, and because we weren't able to make our course under power, we decided to save fuel and set the sails. As seas were calm, we also raised the staysail, hoping that would help us steer a little closer to the wind. At noon, we recorded 90 nm for Day 4.

Throughout the afternoon, *Windy Lady* glided effortlessly through quiet, rolling seas at four kt and we again soaked up warm sunshine. With hardly a cloud in the sky, the sailing was extraordinary, even if we weren't going in the right direction. The magic continued into the next shift, when I again watched the sun drop over the edge of the world, and the night sky fill with brilliant stars.

When Dave relieved me at 2000, the breeze was down to 8 kt but had shifted to the west, so we were now able to make our course. Conditions remained unchanged through his watch and for an hour into mine, but then the winds increased to 15 kt and veered sixty degrees, putting them abaft the beam. For probably the only time ever, I rousted him from his berth, and we dropped the staysail and put a second reef in the main.

Soon after he took the watch at 0400, the winds backed ninety degrees to WSW and increased to 15–20 kt. With winds again on the nose and a five-foot westerly swell, *Windy Lady* was pushed south once more. So at 0800, we tacked, turning the bow northward, and shook a reef out of the mainsail. We were then beating into six-foot seas and heading west-northwest, but after a couple of hours, the winds backed twenty degrees, and we were able to steer straight west. We logged only 85 nm on Day 5, and while we struggled, *Night Music* was advising all boats to stay north of Latitude 30S.

Just after midday, the winds eased to 12–15 kt, and then slowly backed twenty degrees over the next two watches. With rough six-foot seas, we now had to assist Otto, but at least we were going in the right direction. Dave made sixteen nm during his watch, while I made eighteen in mine. He picked up another nineteen nm at midnight, but winds had then dropped to 10 kt and seas were down to three feet.

For the rest of the night, the breeze wandered between

S and SSE at 0–8 kt, and we struggled to keep Windy Lady moving forward. I also spent three hours of my watch looking at the lights of a sailboat as it motored past to the south. By 0800, the air was barely stirring, so Dave reluctantly started the engine. We recorded 86 nm for Day 6 and were just halfway to New Zealand.

During the warm, sunny afternoon that followed, we motored through calm seas with long, high swells that most resembled a rolling mirror. We continued under a night sky filled with stars and then, about midnight, the wind stirred out of the west. After running the engine for over eighteen hours, we raised the sails at 0230. By 0400, the winds were at 12 kt; by noon, they were at 15 kt and seas were up to five feet.

Under a grey, overcast sky, winds now backed to WSW at 18-25 kt, and we spent the next twenty hours beating into five-foot seas. When we started losing some of the westing we'd made during the previous two days, we held the bow close to the wind, trading boat speed for direction. Early the following morning, the winds backed a little more but then they veered; by noon, they had settled at 15 kt from the west.

For Days 7 and 8, we logged 107 nm and 95 nm, but the warnings from *Night Music* had become more urgent. They were cautioning boats not to cross Latitude 30S, as the low-pressure system that had been developing over Australia had started to move across the Tasman Sea.

By now, the repeated warnings, twice a day for five days, had succeeded in reminding me of all the stories I'd heard in Neiafu. My imagination was running riot, with terms like *gale-force wind, knock-down,* and *lost-at-sea* chasing themselves around in my head. So when we approached Latitude 30S that afternoon, I was dreading the coming days. Silently, I stood and watched the GPS as we crossed the line, and took little comfort when Dave confidently observed, "We're only three days away from making landfall."

A few hours later, the winds strengthened to 15–20 kt and the temperature dropped, so we pulled on long pants and shoes for the night watches. With the breeze now steady from the west, we were putting in reefs or shaking them out at every change of watch, trying to get the best boat speed for sea conditions. By noon the next day, winds were gusting up to 25 kt, and seas were over six feet. We then had three reefs in the main and one in the headsail, and recorded 105 nm for Day 9.

After plotting our noon position, I knew I had to do something to calm the turmoil in my head; I needed to know more about the threat posed by the storm. Digging out a pad of graph paper, I plotted the actual position of the low-pressure center in the previous day's forecast, which had the storm front approaching the east coast of Australia. I next added the latest position, which showed the center had moved east over the Tasman Sea to Latitude 39S; it had deepened from 996 to 979 mb. The forecast was now calling for 35-kt winds in an area stretching north 360 nm.

I then added our noon positions for the same two days. When I studied the result, I felt a huge weight drop from my shoulders. It was clear that we would be in the lee of the North Cape before the storm front closed on New Zealand. Yes, we would face stormy seas, but we would avoid the worst of the weather. With that, I was able to shut out the warnings from *Night Music* and concentrated again on the business of sailing.

Conditions now eased slightly, and for the next eight hours, winds held steady at 20 kt with six-foot seas. I was conscious of a vague hope that the storm might be weakening, but for now, everything was going smoothly. *Windy Lady* was handling the conditions well and Otto was providing yeoman's service. Admittedly, we considered making a beeline for the Bay of Islands but stayed the course, as Dave wanted to be a little farther west before turning south.

That evening, the low was still located north of 40S. Its center pressure had dipped to 973 mb, and the forecast was calling for 40-kt winds in an area extending 500 nm north. Now, as we sailed into the night, the winds strengthened and seas grew higher. During Dave's 2000-watch, winds veered from 270 to 300 degrees and increased to 25 kt. When I came on watch at midnight, winds had blown the clouds away, revealing a spectacular night sky, and seas had grown to eight feet. During my watch, winds increased to 30 kt, so I put a second reef in the headsail.

The weather report next morning placed the low-pressure center at Latitude 41S and the pressure at 970 mb; the huge storm was 1200 nm across and winds of 40 kt were forecasted for 600 nm to the north. The storm front, which had moved across the Tasman Sea at a speed of 30 kt, was now approaching the North Island, bringing rain to an area 60 nm in front. When I updated the information on my graph, it highlighted how slowly we were moving compared to the speed of the approaching storm, but there was no doubt we would beat it to the North Island.

The roar of the sea was deafening when I entered the cockpit prior to my 0800-watch, and I could hardly believe my eyes. With a pale-blue sky overhead and brilliant sunshine, the views in every direction were mind-blowing. Long, regular rollers stretched out in front of *Windy Lady*, with the sun glinting off breaking waves, and the wind ripping long skeins of foam from the crests and carrying them away. Winds were registering at 25–35 kt, gusting over 40, and breaking swells were ten feet high.

We now turned the bow to 170 degrees, just slightly east of south, and headed for the Bay of Islands. We seemed to have the perfect angle for the westerly swells, and *Windy Lady* soared across the waves in what would prove to be the most exhilarating ride of our lives. About 1000, the winds backed to 270 degrees and the swells grew higher, stretching up to twelve feet. The troughs in between were steep-walled and deep, and occasionally, a stronger gust would blow the tops off swells, flattening the water, but they quickly built back up again.

With a triple-reefed main and just a bit of headsail, the ride was so smooth that we could have rested easily in our berths but, powered as we were by adrenalin, we stayed in the cockpit. At noon, the Signet recorded 136 nm for Day 10, although the GPS indicated we'd made good only 123 nm. During the afternoon, the winds backed to 240 degrees, remaining constant at 25–35 kt with gusts over 40 kt. Seas were still at twelve feet, and through it all, Otto performed faultlessly.

About midnight, we started to draw into the lee of the North Cape. After over sixteen hours of gale-force conditions, winds then dropped to 20–25 kt and seas to eight feet. In the 24 hours ending at 0400, we logged 162 nm, giving an average speed of 6.7 kt. On our fastest watch, we logged 31 nm for an average speed of 7.75 kt.

The dawn brought our first glimpse of the low hills of the North Cape, by which time the winds were down to 15 kt with three-foot seas. We waited off the channel to the Bay of Islands for better light and started the engine at 0500, then dropped the sails and motored through the entrance. At 0847 on November 20, we rafted up to a charter boat at the Customs Dock in Opua, and cleared in with the authorities. After eleven days at sea, our rhumb line course of 1,083 nm had stretched to 1,192 nm, according to the boat log.

We were able to leave *Windy Lady* at the dock for a half-hour; so, under a grey, stormy sky, we walked down the long pier to the Post Office, where we'd been told there might be information on a berth. After the intensity of the past few days, even that simple task had a dreamlike quality. When Dave then heard someone call his name and turned to see a familiar face standing only a few feet away, he was momentarily bewildered.

The face belonged to a former client, who had stopped in Auckland while traveling from Tahiti back to

Vancouver. With time on his hands, he'd driven up to the Bay of Islands for the day. Unfortunately, Dave couldn't stop for long, as we needed to return and move the boat. When we walked on, however, he pondered over how it was possible that someone from our hometown could be walking up the dock at precisely that moment.

At the Post Office, we learned there were no marinas nearby, but a mooring buoy was available just across the inlet. Having no desire to anchor out, we took it without a second thought. After returning to *Windy Lady*, we motored over to the buoy and settled into what would be a six-month stay in New Zealand.

For the first week, we sat and watched as the rest of the fleet arrived from Tonga. Some fifty sailboats made the passage at that time and most took from two to three weeks, with the shortest time being ten days and the longest twenty-three. A few boats were becalmed, others stopped at Minerva Reefs to wait out weather, and several ran out of fuel and ended up being towed into port at Opua.

CHAPTER 16

Land of the Long White Cloud

Our mooring buoy was just a ten-minute dinghy ride away from the Opua Boat Club. There we had access to a dinghy dock, amenities block, bar/restaurant/lounge, and potable water on the fuel dock. Other than the Post Office, however, the village offered only marine repair and supply facilities, so we had to buy our provisions elsewhere.

On our second day in port, we hiked into the town of Paihia, following a six-km trail that wound beneath the cliffs along the shoreline. The day was pleasant, with scenic views down the channel, and it felt good to get out and walk after so many weeks without exercise. But we couldn't bring back much in our backpacks, so a few days later, we bummed a ride with a local woman into the town of Kerikeri and bought more supplies.

As sailboats arrived in port, we attached faces to the

names and voices that we'd heard for weeks on VHF radio. For yachtees, names actually meant first name and boat name. Amongst couples like ourselves were Pony and Sylvia on *Rassamond*, Ivan and Trish on *Lancia II*, Ken and Pat on *Iron Butterfly*, Del and Joanne on *Limbo*, and Mark and Abee on *Anesthesia*. There was an older couple, Bill and Gail on *Bright Wing*; a family with a young lad, John, Kim and Alex on *Moonshadow*; and the single handers, Skip on *Wild Flower*; Don on *Sourdough*, and Leo, our friend from Tonga, on *Valkyre*. Many of the boats were American, with only a sprinkling of Canadian and European flags.

It was quickly apparent that anyone staying in Opua would need a car, so there was a bit of a rush on used vehicles. When Dave heard one advertised on the cruisers' net, he quickly contacted the owner and was soon driving around in a 1979 Datsun sedan. In the process of acquiring it, we learned a little about the local banking system (transferring funds from Canada), as well as registration and insurance requirements.

Having a car gave us access to shopping in the nearby towns of Kerikeri and Paihia, as well as in Whangerei, a larger town about an hour's drive south. We soon learned that, except for meat, prices were generally more expensive than in Canada. Early in December, we drove south to Auckland, checking out marinas, but didn't see anything we liked better than where we were. Pony and Sylvia, who joined us that day, took a different view and eventually moved *Rassamond* down to the city.

On those first trips, I was dismayed to see hundreds of

dead opossums littering the highway; there were too many to even count. We later learned that 'possums were only one of many species introduced to the country that had no natural enemies and had become problems. Even more sobering were the dozens of white crosses on the roadside that marked traffic fatalities. These memorials didn't seem to cause drivers to reduce speed or improve their driving habits, and we were soon of the opinion that Kiwis were the worst drivers we'd ever seen.

In all fairness, they seemed to accept with equanimity the annual influx of yachtees who cluttered up their roads, especially as many of us had never driven on the left side before. The cars were all right-hand drives, which meant manipulating the gearshift lever with the left hand, while windshield wipers and signal levers were reversed on the steering column. More importantly, the driver's perspective on the space occupied by the vehicle was very different.

Accidents happened and a local woman rear-ended a yachtee when he stopped for traffic at a one-lane bridge on route to Auckland. She jumped out of her car, crying accusingly, "You could have made it!" As she had no insurance, he had to find $600 in his cruising budget to pay for repairs. If he hadn't, he would have lost the Warrant of Fitness, without which he couldn't sell the car. Fortunately, nobody was hurt.

On Christmas Eve, a van disappeared from the boat club parking lot, where all the cruisers left their vehicles. When it turned up a few days later, thieves had tried to strip it of everything of value, including the air-conditioning. Dinghies also went astray occasionally, and it wasn't unusual to find them at the high tide line on the opposite shore.

In mid-December, we noticed that a smaller boat moored next to us was getting a little close; one morning, we woke up to find it only six or seven feet away. At the time, *Windy Lady* was pulled right back on her mooring line, so the other boat was riding up on hers. Our landlord had us swap places with his boat, and a few days later, we watched with interest as a work barge edged up to our old buoy.

After lowering two stabilizing legs, the crew picked up the buoy and winched it aboard, bringing up the end of a very large, heavy chain. The diesel engine then growled loudly as it lifted something very heavy off the sea bottom. Later, we dinghied across and talked to the operator, who told us that he used GPS to find and reposition the weight, which was a two-ton cement block. He also told us that there was a back eddy in that part of the bay, and it seemed to be affecting lighter boats more than usual.

It was comforting to know the buoys were well secured, but we only realized the significance a few days later, when we listened to the morning weather forecast. A cyclone was poised to track down the east side of the North Island and would hit within twenty-four hours. With winds of 80 kt at the center and 60 kt within 180 nm, it would also bring a "phenomenal" swell. Dumbfounded, we looked at each other in disbelief; we thought we'd left the threat of cyclones in the tropics!

The first warning for Cyclone Fergus came on

December 29. According to the report, a high-pressure area had stalled to the east of New Zealand, creating a squash zone between it and the low-pressure area coming behind. This had caused the cyclone, which had been moving southeast between systems, to suddenly turn and head straight south.

We prepared as best we could, adding a second brand-new line to the mooring buoy, and wrapping a longer line around the boom to secure the mainsail and its cover. Everything above and below decks was stowed or tied down, including the dinghy. I then spent the afternoon listening to weather reports and conversations on VHF radio until my stomach was in knots. Other crews were securing their boats; some were also looking for moorage or setting out additional anchors.

The day started with winds of 20 kt, heavy rain and limited visibility. The winds picked up to 30 kt during the afternoon, fluctuated a bit, and by bedtime were again at 30 kt. We decided to keep watch overnight, as winds anywhere close to forecasted speeds could be disastrous. I took the first four-hour shift, and Dave relieved me at 0200, at which time swells were surging down the inlet. He didn't stay up for long however, as winds began to ease; by morning, they were down to 15 kt. The cyclone had moved southeast after all.

Still, the storm caused problems. When the wind shifted, a few boats dragged their anchors, running into other boats and fouling anchor lines. Torrential rain also played havoc with the many campers who spent the Christmas holidays in the Bay of Islands. Campgrounds turned into muddy pits, and long queues formed at the local laundromat as people waited to dry out their gear.

Another cyclone warning was issued ten days later, on January 8. Cyclone Dreena was tracking south-southeast towards the west side of the North Island. With a center pressure of 945 mb, it had the potential to do a lot of damage. By the time it hit on January 10, the threat had lessened; the center pressure had risen to 983 mb and the forecast was then calling for maximum winds of 60 kt.

At 0900 that morning, visibility was poor in heavy rain, with winds gusting to 25 kt. At 1000, gusts were hitting 30 kt; at 1100, they were up to 40 kt. At that time, we went up on deck to check the boats around us. All were straining against their mooring lines, with waves breaking over the bows, but appeared to be riding out the storm safely. Shortly after noon, we recorded a top wind speed of 42 kt. At the long dock across the channel, the three-masted tall ship, *Spirit of New Zealand*, recorded a wind speed of 55 kt.

The winds eased for a few hours mid-afternoon, but were back up to 20 kt by 1800, with gusts up to 35 kt overnight. The gusts, along with heavy rain, persisted well into the following morning, when the weather started to clear. The only damage we heard about was from Leo on *Valkyre*, who reported that a motor launch had passed close by and cut his anchor rode. While he had another anchor ready to deploy, he lost the first one along with 40 feet of chain and 100 feet of nylon rode.

There seemed to be a lot of bad weather that summer, trapping us onboard for up to five days in a row. We would then winch the dinghy up onto the foredeck because waves sweeping down the inlet caused it to buck like a bronco. The bow would rear up almost level with the upper lifeline, six feet above the water. I thought it amazing that the sea simply didn't run in over the transom. When we did leave it alongside, it filled up with rain and we bailed out buckets of water, sometimes two and three times a day.

With hours to fill, we did a lot of reading. Dave also amused himself by tying knots, laboring for hours over a Turk's Head, or baking cinnamon buns, at which he became adept. On one of our first visits to Whangerei, he purchased a transistor radio, and we then discovered talk radio. It seemed to be very popular with the locals and gave us a window on the country that we wouldn't have had otherwise.

Primary topics were national politics and crime, particularly theft, but nuclear testing and nuclear waste were hot button issues. The environment itself didn't seem very high on the agenda, even with a hole in the ozone layer. Lotteries, along with horse and dog racing, seemed to take an inordinate amount of money out of people's pockets, while sports were embedded in the national psyche, including rugby, America's Cup sailing, and any new activity that provided a thrill.

The thrill seeking seemed to be reflected on the nation's highways, with fifteen dead on a three-day weekend in a population of 3.5 million. There were also frequent reports of people in need of rescue as boats capsized, kids drifted out to sea on air mattresses, and hikers got lost. In between, commercials promoted bee pollen, deer velvet, and other home health remedies.

Our only other distraction at the time was a regular morning visit from a family of ducks. The mother always quacked loudly to announce their arrival and, on her first visit, brought six ducklings, one of which was solid brown. Dave then crouched in the dinghy, feeding them oat flakes; soon the brown one, which he named Fred, was sitting in his hand. Two of the ducklings were missing on their second visit, and the last time we saw the family, two more were gone, including Fred.

The reality of taking a dinghy ride every time we stepped off the boat now became tiresome. Even when it wasn't raining, winds blowing over tidal currents often created high waves in the channel between the boat club and our mooring buoy. We started using the hard dinghy again, as it was drier than the inflatable, and with practice, Dave almost had it ferrying across the bigger waves. The dinghy dock would be moaning and groaning as waves surged underneath, and he would then push through two or three rows of heaving boats and motors in order to tie up.

When the weather improved, we started exploring in the car, occasionally just following paved, two-lane roads through the countryside to see where they went. With no shoulder between ditch and pavement, the roads seemed very narrow, while all the bridges had only one lane. Along the way, we visited Haurau Falls and Bledscoe Lookout, with its panoramic view of the Bay of Islands. We also admired ocean views along the Tutukaka Coast, and spent an hour with locals in the park at Whangerei Falls.

We were seldom out of sight of flocks of sheep or herds of cattle, as thousands grazed in fields that once grew gum and kauri forests. Some of the land had since reverted to bush, but the yellow scars of access roads showed up clearly on the ridgelines. I took particular pleasure in the wild flowers that grew alongside the roads, large, showy blossoms that reminded me of domestic plants— and then there were the brilliant red blossoms and dark green foliage of the Kiwi Christmas trees that dotted the seascapes (pohutukawa trees).

One day, on our way to the island's west coast, we saw a sign advertising "The Old Monarchs", so stopped to check them out. I'd read that Kauri trees could be very impressive, with the largest having circumferences of forty to fifty feet at the base. The trunks then split about halfway up, producing multiple tops over 100 feet high. Unfortunately, the few trees we saw that day weren't very inspiring, located as they were in the middle of a scrub forest of willow-like saplings and small trees, with barbwire spread across the ground "to protect feeder roots", and the chatter and clatter of visitors echoing off the boardwalk.

Another day, the car broke down on the highway, leaving us stranded out in the country. Dave pulled off near a gate on a side road, but we had no idea where to go for help, so I started walking north and he stayed with the car. With his usual luck, a farmer arrived at the gate within thirty minutes and offered to phone for a tow truck. A garage happened to be located a few miles to the south, so a mechanic soon arrived and fixed the fuel pump, which was the problem. Dave then realized he didn't have his wallet, but was able to convince the chap that he would return next day and pay the bill.

Meanwhile, I walked for half an hour, self-consciously sticking out my thumb. When a vehicle finally stopped, I turned and peered into a van. I don't know what I expected, but I found myself staring into the face of a very large, very black man. My first instinct was to keep on walking but then I noticed a boy, maybe ten years old, sitting in the passenger's seat, and a whole load of oranges covering the back of the vehicle. In the next moment, I was explaining my predicament to the driver.

As I climbed into the back seat with the oranges, he responded, "I saw the car back a ways with the hood up, so wasn't surprised to see you walking down the road." I then learned that the driver of the vehicle was a recent immigrant. He and his family came from Tonga, and he'd spent the morning picking oranges with his son.

After driving ten km into the next town, he dropped me in front of a garage. I located the mechanic and we returned up the highway in a tow truck. The car was gone by the time we arrived, and we then caught up with Dave at the edge of town. I paid my mechanic $30 for his trouble; next day, we returned and paid Dave's mechanic $40.

While we didn't get a lot of exercise that summer, we did hike the six-km coastal trail into Paihia several times, and twice walked a similar distance along a lightly traveled country road into the small, historic town of Russell. On our second visit, we climbed Flagstaff Hill to watch the Tall Ships' Race. Although a beautiful, sunny day, it was very windy and not that warm, but the views were splendid, with the village nestled in green hills around a small bay and boats sailing down the channel. Other than that, we occasionally climbed the low hill next to the anchorage and visited with Tony and Mary, our landlords, who lived in a very comfortable house overlooking the bay.

When a stretch of good weather settled in, Ivan off *Lancia II* persuaded Dave to enter a Fun Race; the route would take us forty miles up the coast. But when race day arrived, all sunny and warm, there wasn't a breath of wind. Ten minutes after the start signal, racing boats were still sitting on the start line. Five minutes later, there was a roar of engines as cruising boats, which were lined up behind, moved off under full power. Only then did we realize that our race was on, but without wind, it wasn't much of a race.

After motoring all afternoon, we anchored near the Game Fish Club in the harbor at Whangaroa at 1700. Upon going ashore, we strolled along the waterfront near the club where officials were weighing in an incredible-looking sailfish. Weighing in at 210 pounds, the fish was close to nine feet in length from the tip of the bill to the end of the tail fin. We then met up with Ivan and Trish and had a few beers at the club. As none of us was interested in the barbeque on offer, we pooled our resources and returned to *Windy Lady* for a late supper.

Trish brought the mince and Dave cooked spaghetti, and we spent the evening just enjoying each other's company.

Next morning was bright and clear, but we weren't due to return until the following day, so were a little slow getting started. About 0900, I noticed a parade of sailboats leaving the harbor, which struck me as odd. Tuning in the 0930 weather broadcast, we learned that strong, southerly winds were pushing up the coast; they were expected to hit 30–40 kt by evening and would last two to three days. As the harbor was surrounded by high, steep hills, and reportedly very squally in stormy weather, Dave made the decision to return south also.

After we'd winched the dinghy onboard and raised the anchor, Dave took the helm, steering the boat out through the inlet, while I went below and secured the cabin. Once we were clear of the entrance, we raised the sails, setting off under a bright, sunny sky. Within a half hour, winds were on the nose at 20 kt; a few hours later, they were up to 30 kt. Although a few boats tacked far out to sea, we initially took an inshore route, but ended up fighting high swells and strong currents. We tried motoring but barely made two kt, so reset the sails and tacked farther out.

An hour before we reached the Bay of Islands, winds were gusting up to 40 kt and Dave had his hands full at the helm. Before making our last tack into the inlet, we stopped and put a second reef in the main, then partially furled the jib. At 2030, we saw *Lancia II* in the grey twilight; her dinghy was then bouncing along at the end of a towline and turned turtle soon after. We passed the marker buoys off Russell in full darkness, at which point the swell began to ease, but winds continued to howl down the channel.

Somehow, Dave homed in on our mooring buoy, but I couldn't keep it in sight in the surging waters, so wasn't able to hook it with the boathook. After three attempts, we gave up and, at 2315, dropped the anchor nearby. Next morning, we tried again but winds were still gusting to 30 kt, and waves tossed the boat in one direction and the buoy in another. This time, we succeeded on the third try. Strong winds then kept us onboard for the next two days.

On a pleasant summer afternoon towards the end of January, we took *Windy Lady* out for a sail, stopping at the fuel dock on our return to fill water tanks and scrub down the deck. Almost immediately, we noticed a couple of young lads on a sailboat tied up nearby. Maybe nine and twelve years old, the boys were obviously enjoying themselves, climbing and swinging on heavy ropes strung between the pilings next to their boat.

I was always intrigued when I saw families living on sailboats. I knew that most of the children were home schooled but wondered how they managed in such a confined space. As we watched, enjoying the sight of kids just being kids, one of those "boat moments" occurred. The younger lad's glasses slipped from his face, falling into the water. The father's reaction was extraordinary; he dove after them so quickly it seemed he would surely grab them before they hit the surface. Of course, that didn't happen. Upon reappearing, he ruefully explained that they stayed just beyond his fingertips.

Early in February, we packed up the car, planning to do some touring on the South Island. After driving south to Auckland, we headed over to the west coast, arriving in New Plymouth just as the setting sun lit up the perfect cone-shape of Mount Egmont. We thought we would stop wherever the spirit moved us, but soon learned that it was holiday season. If we wanted a room, we had to make reservations.

We stopped next in Wellington, where dark clouds sitting low on the hills brought wind and rain. We spent most of one day at the Maritime Museum, fascinated by the exhibits, especially the story of an inter-island ferry called the *Wahine*. This large, modern vessel ran headlong into tropical cyclone Giselle in April of 1968 and foundered almost within sight of the city. Stories of the desperate efforts made to rescue the 734 passengers and crew were spellbinding.

The next morning we caught a ferry to the South Island. The weather improved as soon as we left Wellington harbor, and seas were flat as we crossed the notorious Cook Strait. By the time we sailed down Queen Charlotte Sound and disembarked at Picton, the afternoon had turned very pleasant. We drove to Nelson and continued around Tasman Bay to Golden Bay, where Abel Tasman came ashore in 1642. (He called it Murderers Bay, after Maoris killed four of his crew.)

Under cloudy skies, we crossed the saddle to Greymouth on the west coast, where there was evidence of serious flooding from torrential rains a few days earlier. The sky cleared as we drove south and, with warm, sunny conditions, we stopped frequently and began to relax. Wide, shallow riverbeds cut the highway periodically, all of them spanned by long, narrow, one-lane bridges. Seascapes changed around every corner, as swells rolling across the Tasman Sea crashed onto the island's west coast. We strolled along sandy beaches and hiked into both Franz Josef and Fox Glaciers, as well as Paparoa National Park. Somewhere along the road, I noticed a forest of superb tall trees, surrounded by heavy undergrowth.

Turning inland at Haast, we drove to Wanaka, where we stopped for the night and stayed a week. Perfect weather turned it into our ideal playground. We climbed through fog to the top of Mount Roy and were enchanted when the mist lifted, revealing spectacular views of Glendhu Bay and Lake Wanaka. We hiked up Mt Aspiring and into Rob Roy Glacier, and swam in the glacial waters of Lake Haura, after hiking the Sawyer Burn Trail. At the end of each day, we sat and barbequed steaks on a grill outside our motel room.

Only an early morning phone call from the manager on a Friday, telling us that the motel, including our room, was fully booked for the weekend, prodded us into moving on. Then, when we stopped for gas on the way out of town, the attendant told Dave we were lucky, because normally the summers were cloudy, windy and rainy.

After reaching the east coast, we stopped at Dunedin and visited a cold and windy albatross sanctuary, where we learned that adult birds practically needed gale force winds to get airborne. At Turtle Rocks, we inspected large, round rocks strewn along a sandy beach; most were about a yard through and their cracked surfaces did resemble turtle shells. As we drove north, most of the eastern plains appeared to be under cultivation. I wondered if they had the same problem as Opua, where fertilizers leached into streams and ended up in the ocean.

A flat tire going into Christchurch on a Saturday resulted in a two-night stay in that beautiful city because we couldn't get it repaired until Monday. The Flower Festival had just ended, so we were able to admire the displays in the downtown area. On Sunday, we toured the Air Force Museum and Antarctic Center, racing to and from the car through driving rain.

After two more nights on the east coast, we caught a fast ferry out of Picton, cutting the travel time across Cook Strait by one-half. Arriving ahead of schedule in Wellington, we couldn't find a place to stay. By 2100, all motels had posted 'No Vacancy' signs, and we ended up spending a very uncomfortable night in the car. Next day, when we started to see signs promoting Rotorua, we heaved tired sighs and kept on driving. In fact, "get-home-itis" had set in and we arrived back in Opua the following afternoon.

We felt very comfortable traveling about New Zealand; not surprising I guess, given our shared history of British democratic institutions and common law. But we were interested in the differences too, and were intrigued to learn that New Zealanders had recently changed their electoral system from first-past-the-post to multi-party-proportional.

Previously, there had been ninety seats in the

legislature, each filled by a member representing voters in an electoral district. As part of electoral reform, the number of seats was increased to 120, but the members directly elected by voters was reduced to sixty-five. The remaining fifty-five seats were assigned to lists candidates, who were selected from lists provided by political parties in advance of an election. The number of seats each party was allocated depended on the percentage of popular vote they won, which was determined by voters in a second ballot.

The first election incorporating these changes was held on October 12, 1996, six weeks before we arrived in the country. No party won enough seats to form the government, and for the next three months, leaders of three political parties haggled over power-sharing arrangements. The leader of the third-place finisher then called a press conference and announced that his party would form a coalition government with the National Party (who found out along with everybody else).

The coalition had only a one-seat majority in the legislature, and according to talk radio, the junior partner (with 13% of the popular vote) ended up with a disproportionate amount of power, dominating government policy. During the six months that we followed events, the shenanigans of the political class took priority over the running of the country, and the government just stumbled from crisis to crisis.

Dave and I talked a lot about this at the time. We both believed in a democratic process where informed citizens held elected members accountable. We couldn't understand why voters would have agreed to turn over 45% of the seats in the legislature to list candidates, who owed their loyalty to political parties, not to voters. Moreover, the allocation of the fifty-five seats wasn't based simply on the popular vote. If so, how did one party with 28% of that vote, and another with 13%, both end up with eleven list seats?

We concluded that a voting system has to be simple and transparent. If the objective is providing stable government, then politicians should be promoting policies that unite people, not divide them. That wasn't what we saw happening in New Zealand. Whatever its problems, the one person/one vote, first-past-the-post tradition, still seemed better than the alternatives. Yes, a voter has only one vote, but in a democracy, that has real power.

CHAPTER 17

A Sailor's Life Ashore

If I thought life on *Windy Lady* was going to be one long adventure, I soon learned otherwise. The sea environment was harsh, with humid, salty conditions exacting a toll far beyond normal wear and tear. While we scheduled many jobs for the annual haul out, there was always work to do, especially before and after a passage. Because our income was limited in those first years of retirement, and Dave had gone well over budget when outfitting the boat, we did most of it ourselves.

In the time we'd been on the boat, we'd learned there was no such thing as *blue* or *pink* jobs (i.e. male or female); just a lot of work that often required two pairs of hands. Still, the responsibility for equipment and boat maintenance fell on Dave's shoulders, so I pitched in wherever I could. Some jobs came easily, like stripping down winches and inspecting fittings and halyards on the foredeck and the mast. But I also volunteered to do the mast work, simply because I wanted him on deck, controlling my tether, rather than the other way round.

We started on our "to do" list as soon as the weather settled after we arrived. The leak in the port cabin had priority and we wrestled the mattress up on deck, so I could scrub down the saltwater stain on its side. He then traced the leak back to a deck drain, not to the porthole as we'd thought, and resealed it. The settee cushions were also damp from a combination of salt spray in the air and on our clothes during the storm, so I bought a spray cleaner and spent hours brushing them, only stopping when my wrists ached.

With salt air driven into every corner of the cabin, we scrubbed down walls and shelves from the head to the stern cabins, and then oiled the teak trim throughout. A season in the tropics had also taken a toll on the bright work, so Dave sanded and varnished the sole in the salon (three coats) and the sides of the companionway (five coats). Meanwhile, I cleaned and re-greased our six winches, drawing sketches so that I'd know how to put them back together.

We resealed the hatch in the galley and replaced fittings in the hydraulic system; I then pulled the anchor chain up on deck and replaced the flagging tape that we were using to mark off twenty-five foot sections. He fixed the strobe light on the masthead, cleaned the hoses and fittings in the head, and spent his spare time trying to free the pin on the beak of the whisker pole, which had frozen shut.

With heavy rain, the rivers emptying into the inlet carried sediment and fertilizers (nitrates) from the fields; at its worst, the water in the bay turned brown and smelled like a barnyard. As barnacles and algae grew rapidly, Dave took the dinghy ashore every three weeks to scrub silt and barnacles off the bottom. Barnacles also plugged the thru-hull on the engine cooling system, a situation we discovered when the engine overheated while we were charging the batteries. *Windy Lady's* hull was always in need of cleaning, too, while the carburetor on the outboard engine gave us nothing but grief.

In between chores, Dave researched two expensive items on his wish list: a RIB (rigid-bottomed inflatable dinghy), and a hard dodger. We spent countless hours visiting boatyards and talking to yachtees, but in the end, he decided there wasn't room in our budget for both. While the RIB would have been nice, the dodger was essential. As he liked the design and workmanship of those built by a local farmer, Warren Patterson, he engaged him to build one. We'd already had dealings with Warren, as he'd sold us the car, and he was prepared to let us help with the grunt work.

On a blustery, grey day towards the end of January, Dave dinghied over to the fuel dock and picked up Warren and his gear, which included a small generator, a skill saw, a tape measure, two sheets of 9 mm plywood and some pieces of timber. With *Windy Lady* fidgeting at the end of her mooring line, the two men started cutting and fitting a framework together. Warren continued to work in spite of 25-kt gusts and lowering clouds. When he was done, we took the boat over to the fuel dock and loaded the structure onto the back of his pickup.

A few days later, we drove out to Warren's farm, prepared to go to work. The pieces were now glued together and sitting in an old shed. Another sheet of plywood was held in clamps nearby; it would form the roof and had been curved to a drop of one inch for every three feet. After sanding the glued surfaces, we began laying fiberglass. Warren cut and placed the matting, Dave applied the resin, and I rolled out the bubbles. Working for most of the day, we finished the outside of the frame and top of the roof; next day, we completed the inside.

The following morning, another grey, windy day, we took *Windy Lady* over to the fuel dock and met Warren for a fitting. The men now positioned the dodger on the cabin roof and set the curved roof on top, so Warren could check it one last time. They then loaded it back on the truck, and we returned the boat to the mooring buoy. We spent the afternoon out at the farm, sanding and putting on a coat of bog (filler). The next day, we sanded off most of the bog, and that was it until we returned from touring the South Island.

When we re-appeared at the farm in early March, the roof was on the dodger and Warren had cut the openings for the windows, reinforcing them with pieces of kauri timber. We worked for the next six days, fiberglassing and bogging, and by the time the dodger was ready for its final fitting, we'd accumulated a total of ninety-three hours.

Despite the threat of Cyclone Gavin, we took the boat

over to the fuel dock and positioned the dodger on the cabin roof. The final adjustments didn't take long, but putting on the "doublers" that would hold it in place was a fiddly job. Later that afternoon, we dropped the dodger off at Doug's Boatyard for painting.

Dave had arranged for *Windy Lady* to be hauled out too, but the boatyard wasn't ready for us when we motored over at high tide. We waited but ran out of daylight, so returned and picked up our mooring buoy in the dark. Next morning was grey and miserable, with steady rain and a super high tide. When we returned to the boatyard, the dock was awash with only the tops of posts visible.

That was the beginning of a very frustrating day. First, the boat didn't fit onto the cradle that was supposed to haul it ashore, then the power washer wouldn't start, and neither did the winch needed to haul it up into the boatyard. By the time the boat was propped up alongside the boathouse, we'd been standing out in the rain for six hours.

Living on the hard was not an experience I enjoyed. With no water, I hauled dishes up and down the ladder after meals and, as much as I tried to keep the decks clean, dirt tracked in, grinding into teak and fiberglass. All the salon lockers were soon open with a muddle of tools and supplies exposed, so our living space was limited to the galley. What made it even worse was the uncomfortable tilt to the deck as *Windy Lady* leaned over on her supports.

Work filled our time for the next twenty-four days, and Dave started by wet sanding the bottom of the hull, covering himself in blue, ablative bottom-paint in the process. He then cleaned the old paint from the propane bottles, as they had started to rust, and applied one that was supposed to be rust inhibiting. Meanwhile, I spent the first six days standing on scaffolding, applying wax to the hull above the water line. It took six coats, all rubbed in by hand, before *Windy Lady* again resembled the boat that left Victoria.

Dave replaced the galvanized fitting he'd put on it Tonga, then the zincs, and cleaned the two impellers. He also replaced the thru-hull beneath the galley sink, which had corroded badly. Doug was now putting primer and two coats of two-pot paint on the dodger, so we sanded it between applications.

Because the deck surface was so slippery, Dave wanted to replace it with a non-skid material and selected a thick paint applied with a looped roller. As it was going to be a big job, he asked Doug to put it on. Still, we did the prep work and spent days on our knees, sanding deck and cockpit, then applied masking tape around windows and deck trim. The day after we finished, Dave was up early, wiping the dew off the deck so Doug could get started. He would repeat that task several times before Doug actually got to work. By the time the deck trim was painted and dry, the masking tape had been on too long; it came off in bits and pieces and was a terrible job to remove. We re-taped the deck before Doug started on the non-skid, but stood by and pulled it off before the paint dried.

Dave also had Doug replace the black Sikaflex that filled the gaps between the narrow teak boards on the benches and sole in the cockpit. He did some prep work first, but Doug did the taping and did a beautiful job of applying the Sikaflex. In the end, we also had him spray on the new bottom paint, as we were running out of time and energy.

When all the work was finished, we still had to install the new dodger. Dave removed a block of halyard brakes that were in the way, as well as the retainers for an old canvas dodger, and filled the screw holes with epoxy. He and Warren then hoisted the dodger up fourteen feet in the air and placed it in position. After attaching it to the coach house roof and doing more bogging, Warren installed an opening hatch in the front. The following morning, a crew from Aqua Glass installed the windows and the dodger was finished, except for touching up the paint.

Shining like a newly minted penny, *Windy Lady* went back in the water that afternoon, but even then, we weren't finished. Ever since leaving Hawaii, Dave had fretted about the house batteries. His fears were confirmed when we returned from the South Island to find them both dead. Giving up on gels, he ordered wet cells to replace them. So now, we tied up at the boatyard dock and made the exchange. Each battery weighed about 130 pounds, so it was a real struggle to move the old ones up the companionway and off the boat, then bring the new ones back onboard. At 1715, after a long and tiring day, we again picked up our mooring buoy.

Three days later, with cruising season just a month away, Dave left for Canada; he would be gone three weeks. I stayed behind because we didn't want to leave

Windy Lady unattended on a mooring buoy. We'd heard too many stories about boats being vandalized; one even had the windlass cut from the foredeck. Besides, there was still a lot of work to do before we left for the tropics.

As I waved goodbye to Dave, my mind was already on the dinghy, and when I returned to the dock, I stood looking down at it dubiously. I knew nothing about the outboard engine and suspected that it was going to cause me trouble, but when I tentatively pulled on the starter cord, it roared to life.

Dave was already unhappy with the new batteries and had instructed me to run the diesel engine for two hours every day and record the charging data. So, as soon as I returned to the boat, I figured I might as well get at it. Engines were included in that list of things mechanical that were a mystery to me but he had left it ready to start, with thru-hull open. I now turned the key, engaging the glow plug momentarily, and it throbbed to life. Other than that, my knowledge was limited to shutting it down with the kill switch.

I was still charging the batteries at 1400, when Doug arrived. He sat in his dinghy, looking up at the dodger, and glumly observed, "That primer coat hadn't been sanded!"

I just stared at him guiltily because, in the rush to prepare for Dave's departure, we'd forgotten. Being Doug, he just shrugged his shoulders and said, "The paint's mixed and won't keep, so where's your sandpaper?"

We set to work and once finished, I cleaned up the dust, while he started to paint. When he was finished, he commented unhappily, "The paint didn't cover the primer very well, so you may want to repaint below the windows in a year or two." He then turned and, looking around the cockpit, touched up a spot in front of the propane locker. Suddenly he froze, staring at the gutter below the seats and cried in disbelief, "Has somebody been grinding metal?"

Uncertainly, I replied, "Uh, Dave did file down the dinghy oarlocks."

"Well, you'd better get busy with Comet, or something similar," he responded icily, pointing to the yellowish-red specks in the gutter and on the coaming.

I spent all the next day on my hands and knees scrubbing the deck and cockpit with comet and scrub brush. As well as the rust specks, I scrubbed off the last of the ground-in dirt from our time on the hard and scraped glue and paint off the coaming and dodger windows with a razor blade. As I worked, I told myself repeatedly, *"No metal grinding or cutting or polishing on this boat ever again!"*

As the afternoon dragged on, the sky suddenly grew darker and I realized rain was not far off. Remembering that the dodger roof was supposed to catch water, I stopped and attached a hose to one of the downspouts and set up a bucket on the side deck. A few minutes later, I heard a scraping noise, accompanied by a gust of wind, and turned in time to see my bucket settle upside down in the bay. Staring at it with actual dislike, I debated where or not to retrieve it and reluctantly decided to do so.

With the blue bottom of the pail moving away

quickly, I gave chase in the dinghy but was unable to grasp it with just one hand. It then took some time to maneuver the boat into a position where I could reach out over the stern with both arms and pull it in. After that, needless to say, I tied the bucket handle to a lifeline. The dodger roof actually proved very effective at collecting water, with each of two downspouts producing a bucket of clean rainwater that night. A few days later, I collected three buckets in half-an-hour.

I spent most of another day working on deck, cleaning paint off cabin windows and deck trim, and washing windows in both cabin and dodger. On the first calm morning, I climbed the mast and spent a half-hour examining halyards, fittings and welds on the steps, looking for anything that seemed different from the last time I'd inspected them. I repaired the jib halyard, which was badly chafed, cutting off almost three feet at the shackle end, and ran the sheets from the mast back to the cockpit, threading them through deck organizers and guides in the new dodger, wondering if they would chafe.

When it rained, I was forced to confront the chaos below decks, where chart table, settee and starboard berth were still piled high. I hadn't been able to put anything away because we'd shoved tools and supplies back into lockers helter-skelter when *Windy Lady* went back in the water. Now, it was simply a question of pulling the lockers apart and re-stowing everything.

I continued charging the batteries daily and started puzzling my way through the brochures we'd accumulated in our battery file, trying to make sense of the data I was recording. When I faxed it down to the dealer in Whangerei, he responded that the batteries were fine and would have to be cycled thirty times before they'd reach maximum capacity.

I had lots to keep me busy, so was grateful to friends who took me away from the boat, asking me over for dinner or meeting me at the boat club on movie night. In particular, there was Len and Pam on *Kapalua II*, Ivan and Trish on *Lancia II* and Ken and Pat on *Iron Butterfly*. Then, one day when I was ashore, I talked to the all-female crew on *Tethys;* they were ready to leave for Australia and had just cleared customs. I also spoke to Skip on *Wild Flower* and he too had cleared, but was waiting for weather before starting the long, non-stop passage to Hawaii. Cruising season had arrived!

When I returned to the dock that day, I pushed the dinghy clear of the other boats and pulled on the starter cord. The engine ran for only a second and then died. I tried several more times, with the same result, so sat and waited for ten minutes, hoping someone would come along. When no one did, I dug out the oars and started rowing.

Winds had picked up while I was ashore, and waves were now sweeping down the channel, so my progress was slow. It didn't help that the oars were short and didn't give much bite to the blades, or that the weight of the engine hanging off the stern caused the bow to sit high in the water. Still, I'd fought my way halfway across when Pam off *Kapalua II* came along and towed me back to *Windy Lady*.

Faced with the prospect of not being able to get ashore, I dug out the operator's manual for the Suzuki. Assuming that the carburetor needed cleaning, I intended to figure out how to do it. But I couldn't lift the engine into the cockpit, so would have to work on it in place. Waves were then jerking the dinghy back against the painter, so I waited awhile, hoping they would ease, but grew impatient.

Climbing down into the dinghy, I carefully removing the engine cover and checked switches, sparkplug, and fuel filters. I was nearly finished when a large wave popped the bow up into the air, sending me face first into the engine. I managed to save myself but was concerned that the next one might put me in the inlet. Finishing up as quickly as I could, I wasn't surprised that the engine still didn't run, as I'd found only a bit of grit in the tank outlet.

As luck would have it, I'd previously arranged to have dinner at the boat club with Len and Pam; they were moored nearby and had offered to pick me up. Over dinner, I told my tale of woe to Len and asked his advice. After listening patiently, he suggested, "If you want to try again, I'll lift the engine onto our deck in the morning and supervise your efforts. But I'm in the middle of my own work," he warned, "so I can't do it for you."

I jumped at the chance, sure that with a little supervision, I could fix the problem. By late afternoon, however, I was regretting that decision. I had taken the engine apart four different times, even stripped the carburetor down to its smallest component, and it still didn't run. I felt frustrated, tired, and guilty for having wasted Len's time.

Len and Pam were going ashore next morning, so gave me a ride in. I then glumly walked over to the outboard engine shop and talked to Charlie, the mechanic. After explaining the problem, I asked, "Do you have time to take a look?"

He responded, "Sure, but you can probably fix it." He then explained, "It's probably the kill switch. They frequently malfunction on Suzuki motors and you don't really need one, because the engine can be stopped with the choke."

Walking me over to a nearby engine, he pointed to three wires coming off the handle and explained how to disconnect the live wire and secure it out of the way. He ended by saying, "If that doesn't work, bring the engine in."

Len again invited me to work on their foredeck but I hesitated, feeling I'd wasted enough of his time. Still, I really didn't want to work on the engine in the water, so compromised and said, "I know you've got things to do, so why not get on with them. If you have time later on, call me on the radio and I'll row across."

Pam called me on the VHF about 1600 and, soon after, the Suzuki was again sitting on their foredeck. It took me only a couple of minutes to remove the cover, find the wires, disconnect the hot lead and wrap it with black electrical tape, then put it all back together. We lowered it back onto the dinghy and, with some trepidation, I pulled on the starter cord—*and it ran*!

I was so relieved and so pleased that, for the first time ever, I just took the dinghy out for a spin. I continued to celebrate for the next few days, going for rides just for fun. Then the engine quit again, leaving me adrift in the middle of the bay. This time it was my own fault, as I'd read in the manual that the fuel switch should be turned off when the engine wasn't in use, and I forgot to switch it back on.

As I sat there, cursing my own stupidity and pulling on the starter cord, I saw a dinghy approaching. When I saw that it was Len and Pam, I came crashing down from my high; I would actually have preferred to sit there all afternoon, rather than be rescued yet again. Len quickly pointed out, "You've flooded the engine, so hold the throttle open while you pull on the cord."

I followed his instructions and the engine started immediately, leaving me feeling even more inept. Trying to hide my humiliation and embarrassment, I stammered, "You truly are a knight in shining armor, coming to my rescue once again." Regrettably, that was not a very good thing to say to someone else's husband. Pam took offense and that was the end of that relationship.

I spent the last week canning another thirty-six jars of meat and vegetables, catching a bus into Kerikeri and then hiking twice into Paihia for supplies. The pressure cooker held only six jars, and each batch took up most of the day to prepare because the stove took a long time to boil water, and I couldn't use the galley sink because the thru-hull was leaking.

With Dave due back the next day, I finally ran out

of things to do. But after running the engine for the last time, I remembered the Velcro strips we'd bought for the settee cushions. Sewing one side to the cushion backs and screwing the opposing side to the wall, my hope was to keep them in place during the next passage. When I sat back and looked around, the cabin interior sparkled almost as much as the outside had, when we put *Windy Lady* in the water three weeks earlier.

CHAPTER 18

The Second Season

By the end of April, a few boats had departed for the tropics, and many others were chomping at the bit. Dave and I worked feverishly at our own preparations, but a couple of days after returning from Canada, he broke a tooth. Needing both a root canal and crown, he scheduled appointments that spanned the month of May.

So, on May 2, we stood by and watched enviously as a small flotilla of a dozen boats departed. Within a few days, however, several had returned to port. That set a pattern for the month, as some boats seemed to return from every group that left; one yacht started out five different times. The problem was that the weather was still unsettled with cyclones developing near Fiji, and strong winds, or winds on the nose, closer to New Zealand. Those boats with schedules to keep pressed onward, but others saw no reason to battle the seas.

Rumors circulated about the difficulties yachtees were encountering, with most stories originating far away on HAM networks. As many boats leaving from Opua checked in with Des at Russell Radio, on a local HF frequency, we started listening in. Des called a roll twice a day and, at his busiest, spoke to more than forty boats. He first transmitted the area forecast, then spoke to each boat individually, sometimes taking an hour to go through the list, as he noted positions and ensured that all was well.

Listening in from shore, we came to appreciate his knowledge and commitment far more than we had on our way to New Zealand, when we tuned in for just a few minutes a day. We heard skippers, obviously feeling stressed by the situation they were in, asking for advice. Des would carefully provide as much information as he could, taking whatever time was needed, and always ended by saying, "The decision is yours." For a boat in difficulty, his voice was a lifeline, always reassuring but never sugarcoated. (Two years later, we had personal experience of that.)

On May 9, boats at sea had been battling 30-kt winds and pounding seas for five days. Crews were tired, and that night, we heard relief in voices that reported winds were starting to abate. The skipper of *Kemo Sabe* was one of those voices, but his day was about to be turned upside down. Des had a message for him from Search and Rescue; they wanted him to turn around and go to the assistance of a yacht in trouble. Apparently, an emergency locator beacon (EPIRB) had gone off that morning from a position some 60 nm south of the yacht's location.

To the skipper's credit, he took only a few minutes before agreeing to turn back. However, he stipulated that he could only do so with a fuel drop, as the boat would be punching into wind and he would have to motor. This was not a surprising request, given that many sailboats had small tanks, so lashed jugs of fuel on deck to provide a margin of safety during a crossing. At the end of the broadcast, Des confirmed that the authorities had agreed to a fuel drop, so *Kemo Sabe* turned around.

When the captain called in twelve hours later, the boat had made maybe twenty nm. Des now had to relay another message, as the promised fuel drop had been cancelled. We listened to the conversation in shocked disbelief, at the end of which the captain again turned his boat around. Opua was abuzz with gossip for days as people tried to learn what had happened. Several months later, in Vanuatu, we met George and Sarah on *Kemo Sabe* and learned the rest of the story.

New Zealand Search & Rescue had picked up an EPIRB signal from a location southwest of Fiji. An aircraft was sent to check it out, and the crew located a sailboat and saw someone wave, but weren't able to determine whether help was needed. The authorities then contacted Russell Radio, asked if any yachts were in the area, then left Des to deliver the message.

When the emergency signal suddenly went silent the following morning, another aircraft flew to the area. This time, the crew dropped a VHF radio. The skipper, a single-hander, reported that his boat had been knocked down and rolled, taking on water. He was fine, but the rigging was damaged and his radio didn't work. He had switched off the EPIRB once he had the engine running and was then on his way to Minerva Reefs.

I found these events very unsettling and couldn't help but wonder how we would have responded in similar circumstances. Firstly, just finding a disabled boat at that distance, in rough seas, would have been miraculous. Secondly, we were untrained in search and rescue techniques, and sea rescues could be hazardous. So, presuming we actually found the boat, how much help would we have been? Reneging on the fuel drop had then made a bad situation worse, and I concluded that a second boat had been put at risk, with little chance of accomplishing anything.

Dave and I had already faced the grim reality that with only two people onboard, neither of us would be able to rescue the other from rough seas. Privately we joked, only semi-humorously, about throwing a Canadian flag into the sea if such a thing happened. We also believed, as apparently did this sailor, that in the middle of the ocean we were on our own.

That message had been reinforced when we crossed the Pacific the previous year. HF weather broadcasts had included a *Notice to Ships*, advising anyone near a given location to be on the lookout for a yacht in distress. Given how few ships we'd seen at sea, we didn't think that message would have been much help at all. (Although it seemed a high tech racing boat could bring the Australian Navy steaming 1500 nm into the Great Southern Ocean.)

The whole issue of safety at sea was problematic because of mandatory safety regulations that entrenched the safety industry in marine activities. With a Canadian-registered vessel, we spent thousands of dollars complying with these regulations and purchased a life raft, emergency beacon, and flares. All of them were time dated, so required servicing or replacing, and we never used any of them.

We bought the EPIRB before leaving Canada; it was the newest version and identified the boat that carried it. It required servicing a year later, while we were in New Zealand, but we couldn't obtain replacement parts because the company had gone out of business. We then bought a Kiwi-made unit for one-fifth the cost; although it didn't identify the boat, it provided exactly the same chance of rescue.

The issue of safety regulations was one that reportedly caused some yachtees to bypass New Zealand in the first years after the Queen's Birthday storm. The Kiwis introduced new regulations, including mandatory inspections, which had to be met before a departure clearance would be issued. Many overseas cruisers felt they had complied with the safety codes of their own countries, and that such additional regulations were onerous and costly, so some just didn't go to New Zealand.

The longer we were on the boat, the more convinced we became that our safety depended on staying onboard in the first place; anything else was a waste of time and money. That belief was supported by numerous reports of yachts that survived intact after being abandoned by their crews; one even drifted 2,700 nm to the Seychelles. A suggestion then bandied about was to get into your life raft only if you had to step up out of your yacht. (I later read an intriguing news story about a sailboat that drifted away from its mooring on the west coast of Australia and, all on its own, sailed to South Africa.)

Still, I was astonished when we met a cruiser who didn't carry a life raft, but relied solely on a dinghy. In later years, however, I found myself looking at the space taken up by the raft on the floor of the port cabin, wondering whether we would bother to have it repacked when it came due.

As the days between dental appointments dragged slowly by, Dave found a few more jobs to do. After fixing the leak in the galley plumbing, he removed the hydraulic pump at the inside steering station and the salt-water pump on the engine, then had them both rebuilt. He re-installed the block of halyard brakes inside the dodger, then added reading lights, and re-fastened the modified handrails to the cabin roof.

We continued charging the batteries daily and, on May 20, ran a capacity test, placing a charge of 20 amps on the batteries until they dropped to 10.5 volts. The test results came in at 325 amps, nowhere near the 480-amp rating. All the data we had submitted now produced results, and we went from talking to the dealer in Whangerei to the supplier in Auckland. The company agreed to replace the batteries, stating that they appeared not to have been properly "formed" before being shipped.

However, Dave was now having second thoughts about lead-acid batteries. They released a gas while they were charging, which set off the shrill propane alarm in the engine compartment. He'd finally disconnected it.

But that caused him to wonder what effect the corrosive emissions might have on the GPS and EMON, both of which were closer. As new gel-cells would cost an extra $600 and the old ones hadn't lasted much over a year, he now dithered. Finally, after pulling out and studying all the information we'd accumulated on batteries and EMON, he decided to give the gels a second chance.

Boats were now starting to make landfall: Leo on *Valkyre* checked in at Cairns, Australia on May 15; Skip on *Wild Flower* made landfall in Hawaii on the 20th. Two boats also arrived in Fiji, which seemed to cause a flurry of activity in Opua, with at least thirty boats added to Russell Radio's roster in a three-day period.

On May 20, a cruisers' net on HF radio reported two acts of piracy in Papua New Guinea. One involved an American yacht in the Louisiades; the other, a fishing boat near Bougainville, where a man was killed. Other reports came in of a boat abandoned and three crewmembers airlifted to Fiji, while another had lost its rudder and was making its way north under sail. Three days later an American boat named *Ora* went up on North Minerva Reef; the boat was lost but yachtees anchored inside the lagoon rescued the crew.

Beautiful fall weather now brought cool mornings and warm, sunny afternoons, so we spent three days cruising in the Bay of Islands, checking out equipment and brushing up on our anchoring skills. As often happened, there wasn't enough wind to test anything, but we did confirm that the dodger roof wouldn't interfere with the movement of the wind vane on the self-steering. We also checked out the sails and greased the slides on the main sail track.

Dave had placed a "For Sale" sign in the car window early in the month but then stewed as the days went by with no inquiries. It didn't help when he talked to other cruisers who were selling vehicles, as most were taking a bath. After two weeks, he received a phone call and that was all it took. He agreed to renew the Warrant of Fitness and pay the registration for six months, and we were able to keep the car until the end of the month.

We made good use of the car during the last week, filling propane bottles and buying supplies, so once again, *Windy Lady's* lockers bulged. We met with Des from Russell Radio, locating his house by the antennas poking up off the roof, and registered for our upcoming passage. On the last day of the month, we made a day trip to Ninety Mile Beach and watched the waves rolling in across the Tasman Sea from Australia. Still, Dave heaved a sigh of relief when he turned the vehicle over to the new owner and pocketed the cash. He'd lost $450 but figured that a fair price to pay for the use of a car for six months.

We were now ready to leave but awaited delivery of the weather fax that we'd ordered from West Marine. I didn't want a repeat of the events we'd had coming into New Zealand and figured it was the solution; however, Dave took some convincing and didn't order it until May 13. The shipment was initially delayed two weeks, which shouldn't have been a problem, but then we heard nothing. On June 3, we were notified that it had been shipped on May 30, so assumed it would arrive any day.

At the time, we were still monitoring Russell Radio. While many boats had safely reached their destinations by then, a few were still at sea. As the anniversary of the Queen's Birthday Storm approached, the weather at sea deteriorated and so did tempers ashore. First, a second radio station tried to broadcast on the same frequency and at the same time as Russell Radio. Then, an overseas cruiser called in while Des was reading the weather forecast, and tried to give his own weather briefing. He even broke into the roll call with advice for specific boats, warning them that a "bomb" was heading in their direction. When the weather improved after a couple of days and no "bomb" developed, broadcasts returned to normal. I didn't envy the boats at sea, knowing how upset I'd been by such rhetoric the previous year.

On June 4, we arranged for the Safety Inspector to do our mandatory inspection. After poking about the boat for half an hour and asking many questions, the officer signed our report. The following day, we tied up at the dock and took delivery of 450 liters of diesel fuel (*Windy Lady* had two large tanks). We also exchanged batteries, as the dealer in Whangerei had sent up two new gel-cells. Dave now prepared a chart of "Do Not Exceed" charging rates for higher temperatures and reprogrammed the EMON.

Although the weather fax still hadn't arrived, we planned to leave as soon as it did, so cleared with Customs. The weather then deteriorated, bringing NE winds, with heavy overcast and rain. On June 9, we decided to wait no longer and arranged for Doug at the boatyard to forward the unit to Fiji. Before the next break in the weather, however, Customs demanded we check back in. We then motored over to the Sea Life lift and paid out $80 to have the bottom power washed, as a lot of sediment had accumulated over the past two months.

Since the first of the month, overnight temperatures had been cool, down to fifteen degrees inside the cabin, with heavy condensation on the windows. The morning of June 11 was even colder, and Dave started the furnace for the first time since leaving Canada. By noon, the temperature inside had risen six degrees, but there still wasn't enough light to read by. The forecast that day reported that Cyclone Kelly was moving southwest.

When we awoke to a bright, sunny sky the next day, we hurriedly moved *Windy Lady* over to the Customs Dock and obtained another clearance. We were soon motoring down the channel for the last time, and as I studied the picturesque inlet, I thought of the people who had touched our lives while were there.

Colleen at the Post Office had been a treasure, providing information to newcomers with endless patience and good humor. We had enjoyed visiting our landlords, Tony and Mary Watson, who had welcomed us into their home. While working on the dodger, we came to know Warren and Jillian Patterson, who gave us a different perspective on New Zealand. Contact with others had been briefer, like the Tongan who'd given me a ride, but the memory of many kindnesses has endured.

CHAPTER 19

Passage to Fiji

We left the dock at Opua at 1145 on June 12; two hours later, we were clear of the inlet with batteries fully charged. The distance to Suva, on the island of Viti Levu, was 1,050 nm on a rhumb line course, so we figured on ten days at sea. Our route would take us a bit east of due north, but we suspected SE trade winds might push us westward once we got closer to Fiji, so hoped to make some easting early in the passage.

Conditions really couldn't have been better for the start of a voyage, with a bright, sunny sky, winds out of the SW at 12 kt, and a two-foot swell. I stood at the helm, looking around in admiration, while Dave went below to shut down the engine. After I'd settled *Windy Lady* on course, he attached the vivid, red sail to the self-steering. Once Otto took over the steering, with the sail waving back and forth, we were free to relax and soak up the sunshine.

Thankful that the long wait was over, I could feel

the stresses of the past weeks dropping away. When the hills of the North Cape started to fade on the horizon, I realized that, unlike the previous year, I had no qualms about leaving land behind. However, with a definite nip to the wind, the temperature cooled off quickly after the sun went down. By change of watch at 2000, SSW winds were blowing straight down into the cabin, and we closed the companionway hatch for only the second time. We both pulled on cruising suits for the night watches, too.

About 2200, Dave saw the lights of a ship in the darkness and tracked it on radar as it passed two nm to starboard; an hour later, a second ship passed at four nm. Then, just before midnight, a slim, crescent moon slipped down into the sea, leaving the heavens studded with brilliant stars. Throughout the night, the winds stayed steady at 10–15 kt but they moved continually, veering forty-five degrees from SSW to WSW.

With clear skies, the morning brought a long, slow dawn, and just after sunrise, the winds began to back. By 0800, they were under 10 kt from the SW, so we turned the bow to 045 degrees as we tried to make some easting. With winds behind us, the mainsail blanketed the headsail, causing it to flog, so we furled it in and sailed with only the main. Three hours later, the winds shifted to the south and we were able to reset the jib. At noon, after twenty-four hours at sea (Day 1), we logged 124 nm.

The winds now backed to SSE and steadied at 10–12 kt, so Dave set the sails wing-on-wing. Under a clear, bright sky, with winds filling the sails on either side of the bow, the boat rolled comfortably through three-foot seas for hours. Just before nightfall, winds moved to the south and gusts of 15 kt began overpowering Otto, so we reset the sails to their normal configuration.

The night watches were again long and cold, but at midnight, with winds below 10 kt, *Windy Lady* was making five kt as she cut through two-foot swells. With an easy, regular rolling motion, we should have slept well but didn't. I suspected the excitement of being at sea just kept me awake until I was exhausted. The winds backed with the rising of the sun and then settled at 10 kt from the SE. A few hours later, they were gusting from 10–20 kt out of the SSE and swells were up to six feet. When I plotted our noon position, we'd gained 59 nm of easting. For Day 2, we registered 123 nm.

Unbeknownst to us, the winds now started working their way around the compass; they would go all the way around before we reached landfall. By midafternoon, east winds gusting at 10–15 kt; *Windy Lady* was then on a beam reach, being propelled straight north into five-foot seas. Gusts up to 25 kt overpowered Otto during my sunset watch, but a reef in the headsail solved the problem until midnight. Winds of 15–20 kt then had us babysitting Otto for the next eight hours.

First light brought an overcast sky with stormy sunrise, and then the breeze began to ease. As the batteries needed charging, Dave ran the engine for four hours, shutting it down at noon. When we then set the sails, winds were down to 6-8 kt and swells were over six feet. We recorded 133 nm for Day 3.

A series of squalls brought rain showers during the

next two watches, with gusts of 15 kt in the afternoon and 25 kt in the evening. Through it all, Otto worked flawlessly. The winds then settled at 15–20 kt for most of the following watch, and when Dave filled in the journey log at midnight, he described his shift as "a sleigh ride", noting that he hadn't touched the helm once.

But shortly after midnight, Otto was completely overwhelmed, and I then stood at the helm and hand-steered. Winds had moved forward of the beam for the first time, gusting up to 30 kt as they backed to the NE. The boat was then racing through a rainy, pitch-black night at six kt. Dave also hand-steered through the following watch, as squally conditions continued and winds veered to ENE.

It was raining when I came on watch at 0800, with winds gusting from 15–25 kt and rough six-foot seas. As *Windy Lady* was then charging off in all directions, we put a second reef in the mainsail, and then a third. Still, we made 25 nm during that watch. When I plotted our position at noon, we'd used up 37 nm of easting, and logged 142 nm for Day 4.

The winds took a breather early that afternoon, so Dave optimistically shook two reefs out of the mainsail. Two hours later, winds returned at 15–25 kt, still from the ENE, and *Windy Lady* tossed and twisted as she plunged into eight-foot swells. The windows started leaking again, with saltwater widely dispersed by the erratic movements of the boat. We also closed the thru-hull in the head, as seawater pushed through it, filling up the toilet bowl, which then overflowed.

The wild, wet ride continued into my evening watch, and just after sunset, I spotted a ship on the horizon. I watched for thirty minutes as it crossed in front of us, the lights appearing and disappearing as we rode up and down on the swells. We put a second reef in the main when Dave relieved me at 2000, which proved fortunate, because winds were howling out of the NE at 20–30 kt by midnight.

Windy Lady was now beating into swells that were over eight feet high, and for the rest of the night, squall after squall swept down on us out of the darkness. The watches were long, cold, and wet, and the boat took a pretty good pounding. Although we fought to keep the bow pointed north, wind and sea combined to push us westward; by daybreak, we'd lost all our easting. At that point, I had to admit that the hard dodger was one of Dave's better ideas. While the roofline was still directly in my line of vision, which had been my main objection, it kept a lot of rain and spray out of the cockpit.

Midway through the morning, winds backed to the NNE at 20–25 kt. I now held the bow as close to the wind as possible, struggling to steer 350 degrees, with *Windy Lady* bouncing off the waves. At noon, we recorded 124 nm for Day 5. Winds held throughout the afternoon, but the sky grew darker and the sea started to ease. At 1700, the wind suddenly died and the clouds opened up, releasing a torrent of rain that lasted over an hour.

As soon as the rain eased, a breeze rose out of the north at 12 kt. Although seas were down to three-feet, the best I could then steer was 300 degrees. We tacked but couldn't make any progress in an easterly direction either, so turned back at 2000 when Dave came on watch. The first in another series of squalls hit out of the NNE just before midnight, bringing gusts of 20 kt and periods of rain. Our best efforts then had us steering 320 degrees.

As the night progressed, we put in reefs and took them out, and occasionally even caught a glimpse of the stars. We were both very tired; it was impossible to sleep with all the rocking, rolling, and banging going on as the boat crashed into waves. By change of watch at 0400, winds were again out of the north at 12–20 kt. At 0800, they were steady at 20 kt and seas were back up to five feet. We were then steering 330 degrees, but Dave was starting to wonder whether we'd make Suva.

During the morning, winds increased to 20–30 kt and seas were soon back up to eight feet. When I plotted our noon position on the chart, our track showed a definite dogleg to the west. Dave was now talking darkly about making landfall in Vanuatu. For Day 6, we recorded only 89 nm; we were then 59 nm west of our course line.

The winds finally resumed their journey that afternoon, backing to the NNW and settling at 20 kt. We tacked, turning to a heading of 045 degrees, and hope revived that we still might make Suva. A few hours later, the breeze dropped to 12 kt and we pulled the reefs out of the headsail. Then, just after 2000, a warm front passed overhead, bringing a wind shift of 115 degrees. The winds, now from the SW, dropped to 7-8 kt in light rain.

With winds light and variable, *Windy Lady* rolled in six-foot seas and Otto couldn't cope. We hand-steered throughout the night, struggling to keep wind in the sails and the boat moving in the right direction. Just after 0800 next morning, the winds strengthened to 10–20 kt, blowing the clouds away and pushing up very rough seas. After we put a second reef in the mainsail, Otto went back to work. By noon, we'd recovered much of the easting we'd lost and recorded 104 nm for Day 7.

As the batteries needed charging, Dave ran the engine for four hours during his afternoon watch. When we started sailing again at 1600, the breeze was steady at 10 kt from the south, with rolling three-foot seas. Deciding that conditions were right to go wing-on-wing, he poled the headsail out to starboard and eased the main out to port. Soon *Windy Lady* was making five kt, rocking gently from side to side as she cut smoothly through the waves.

I spent most of that watch sitting in the rear of the cockpit, swaying with the motion of the boat. With water swishing softly against the side of the hull, I found myself relaxing for the first time in days. I then watched, entranced, as the sun slipped down a cloudless sky and into the sea; minutes later, a magnificent full moon peeked over the eastern horizon and quickly climbed up into the night sky. By the end of my watch, winds were up to 10-15 kt. Our magic carpet ride then continued for another eight hours, while moonlight reflecting off the water turned night into day.

A following sea started to build after midnight. By the time we crossed the Tropic of Capricorn (Latitude 22S) two hours later, swells were up to seven feet and Otto needed help. At 0600, the winds backed to the SE, at which point Dave reset the sails and tacked. With winds steady at 10–12 kt, we shook the second reef out of the mainsail at 0800. At noon, we recorded 117 nm for Day 8.

Southeast winds brought scattered clouds and warm, dry conditions, but they grew stronger during the afternoon and seas started to build. When Dave attempted to deploy the whisker pole, the car on the mast track broke. Unable to either use the pole or lock it back into position on the mast, he tied it down alongside the toe rail.

When I came on watch at 1600, winds were gusting from 10–20 kt and seas were about five feet. Four hours later, I'd put a second reef in the jib, as winds had increased to 15–25 kt and seas were up to ten feet. At midnight, the boat was tossing and twisting erratically in rough twelve-foot seas, so we put another reef in each sail. It was then a challenge to move about, both inside and out, so we were mindful of the climber's rule and always kept three secure points of contact.

Winds backed to the east just before 0400, gusting from 10–25 kt, but seas now dropped to six feet. Daylight again brought bright sunshine and a few scattered clouds. At 0850, we spotted the southern-most headland of Kandavu Island. It was still forty nm away and emerged as a lump, barely visible above the rough surface of the ocean. As it slowly grew on the horizon, conditions abated somewhat. For Day 9, we logged 128 nm.

Winds strengthened in the early afternoon, and gusts of 30 kt quickly pushed the seas back up to ten feet. When we reached the island at 1600, we had three reefs in each sail, and *Windy Lady* was racing through the waves at six kt. After rounding the island and starting up Kandavu Passage, Dave went down to the chart table and calculated the distance to Suva; we didn't want to arrive in the middle of the night.

Deciding that we needed to slow the boat down, he eased out the mainsail and I furled the jib to a fourth reefing point. (I figured three feet along the foot of the jib equaled one reef.) When the boat speed dropped to four kt, he went below to rest. As soon as we drew into calmer waters in the lee of the island, however, the boat speed picked up to five and one-half kt. So at change of watch at 2000, we dropped the mainsail and furled in all but three feet of the jib, again slowing the boat to four kt.

When I returned to the cockpit at midnight, I was feeling very tired. Rough seas and erratic boat movements had kept us both awake for most of the last twenty-four hours. Of course, a few items hammering back and forth in the lockers hadn't helped. But this was my last shift, so I concentrated on watching for traffic and keeping *Windy Lady* positioned in the center of the channel, well away from any reefs.

Before long, we emerged from sheltered waters, and ESE winds were soon gusting at 15–25 kt in seven-foot swells. When I next looked at the knot meter, it was like a glass of cold water in the face; the boat was now making six and one-half kt. I could do nothing to slow it down, as the little bit of headsail still up was needed for steerage. For the next half hour, winds gusted from 30–36 kt, and *Windy Lady* raced through the waves, surfing down southerly swells and rolling in easterly wind waves.

I heaved a sigh of relief when the winds finally dropped to 15–25 kt, but the night was not yet over. Twice in the next three hours, a lull in the storm dropped the winds below 10 kt. *Windy Lady* then wallowed in the swells and I barely had steerage. When Dave came on deck at 0400, he started the engine, wanting to charge the batteries; we both stayed on watch as we made our way into harbor.

We arrived off Suva just at daybreak, with the rising sun shining in our eyes as it streamed through a break in the dark clouds sitting over the island. We couldn't see directly ahead, so anxiously studied the surrounding waters. Eventually, we spotted the leads and followed them through the reef, dropping anchor at 0819 on Sunday, June 22. We would again sit in quarantine for a day, waiting for Customs to open.

After putting the boat to bed (flaking and covering the mainsail, coiling lines, putting equipment away), we sat out in the cockpit, relaxing and discussing the passage. We were relieved to be safe and secure in port. During ten days at sea, we logged 1179 nm, for an average speed of five kt. Seas had been rough, which wasn't surprising, but the weather hadn't warmed up until we were halfway across. I'd worn my winter underwear under my cruising suit and then under my raingear. I'd worn the raingear even when the rain stopped because of spray. The dodger had been well christened, and I had no doubt that the passage would have been most uncomfortable without it.

The following excerpt from Dave's personal journal

best sums up his thoughts: "A year ago, I would have dreaded it and just wanted it to be over. Now it wasn't that bad. You put in your watch, do your chores, and try to get some sleep. There isn't the same concern over bad weather because you've had some and coped reasonably well. While it's a bit uncomfortable at times, you tell yourself that this too will end—and it does."

CHAPTER 20

A Month in Suva

On our way into Suva harbor that morning, Dave noticed that both tach and alternator were not working. The thought that the alternator might fail at sea had never occurred to him before, so we didn't carry a spare. He had time for only a cursory look before shutting down the engine, and worriedly decided that he'd have to check out the entire charging system as soon as he caught up on some sleep.

The next morning, instead of taking *Windy Lady* over to the main dock to clear in with the authorities, he called the Customs office on the radio. Explaining the problem with the engine, he requested permission to come alongside in the dinghy. It was a long shot, as we knew the officers liked to look inside the boat, but he was lucky and found a sympathetic ear. That made me happy, as I'd been steeling myself for a bout with the wharf's cement wall.

We quickly winched the dingy over the side and lowered the outboard engine, then Dave grabbed passports and boat papers and headed over to the wharf. When he returned two hours later, he described filling in forms, all in triplicate, with no carbon paper. The paperwork had seemed endless as he dealt with Health, Immigration, Customs and Port Authority.

Now free to go ashore, we left the boat in the Quarantine Area and dinghied over to the Royal Suva Yacht Club (RSYC), about a fifteen-minute ride away. Dave tied up to a small dock out front, and we then climbed up a few stairs leading into the grounds. A minute later, we were inside the clubroom, which appeared to be a large, enclosed veranda. We would spend considerable time here in the coming weeks, as its comfortable chairs and big, overhead fans made it a gathering spot for yachtees, a place to renew old friendships and make new ones over a jug of beer.

We bought club memberships at the office and paid $30/week for the use of the dinghy dock, clubroom, laundry room, and ablution block. Mail and fax service were also available at the office, and a restaurant and bar were located on the premises. As our immediate need then was local currency, we set off walking. Thirty minutes later, we were in the busy financial district, where we searched for an ATM with a cirrus logo.

Continuing on to King's Wharf, we found our way to the Port Authority office and paid our port fees. Inadvertently, we then left the dock area through a gate used by cruise ship passengers. Three Fijian men were standing just outside and rushed over, greeting us effusively. As they seemed to be welcoming us to Suva, we assumed they were associated with the cruise lines, so quickly explained that we weren't off a ship. That didn't matter; we were tourists!

Shaking our hands and asking our names, the men tried to engage us in conversation. One then shoved two carved masks into our hands and did his best to keep us distracted, while the other two men stepped back and pulled out long, paper-wrapped packages. I just happened to glance past Dave's shoulder as one of the men put a small knife to a piece of wood sticking up out of the paper.

That was when the penny dropped, and I turned away, screeching, "No! No! No!" I fled across the narrow, deserted lane with Dave right behind me. In fact, these men were con artists known as sword carvers, and we'd read warnings about them in our cruising guide. Once they carved your name into the handle, the men would insist on payment, sometimes not very pleasantly.

Next day, after another eight hours sleep, Dave dug out the information we'd accumulated on the charging system. After studying diagrams, he checked connections and performed self-tests, but found only what he thought might be a loose field ground on the high-output 105-amp alternator. Carefully re-fastening the wire, he then started the engine and monitored the charge going into the batteries. As everything seemed to be working normally, we raised the anchor and moved the boat out of the Quarantine Area.

We now took a quick tour of the anchorage, looking for the right spot to call home. But as we neared several catamarans at the head of the bay, the depth gauge quit working and soon *Windy Lady* was aground. Fortunately, the bottom was soft mud and Dave was able to back her out into deeper water. Referring then to a chart, we dropped the hook near a shoal just off the RSYC.

Dave spent a lot of time over the next few days dealing with equipment problems. He first contacted Signet in the US about the depth gauge and was told that it would have to be sent to them for assessment. That was a non-starter because the process would take a month, even airmail, and we didn't plan to stay in Suva that long. He then contacted a marine store in Canada and ordered a spare alternator, but when weeks passed by without a word, we just forgot about it.

He also tried to send a fax to Hydrovane in the UK, ordering spare parts, but the fax wouldn't go through. After trying several times over a couple of days, he phoned them. Of course, with a twelve-hour time difference, that meant taking the dinghy ashore after dark. After placing the order, he complained about the problems he'd had with the fax. The fellow at the other end just laughed it off, saying, "Oh, we only turn the machine on during office hours!"

The weather fax, which should have been waiting for us, hadn't arrived. Doug, at the boatyard in Opua, didn't know where it was either, but offered to contact West Marine and have the unit sent directly to us. We never knew why it was delayed, but the notion that we could still be waiting for it in New Zealand was quite upsetting. Our mail was also missing; it had been airmailed from Canada ten days earlier.

The good news was that the weather was very pleasant. Almost perfect in fact, when compared to the heat and humidity in Hilo, or the cooler temperatures in New Zealand just before we left. A layer of cloud sitting over the island seemed responsible for the moderate temperatures, as I noticed my arms burning when the sun infrequently found a hole in the clouds. Rain showers were generally light, although one night I did collect two buckets of water off the dodger roof.

We were soon exploring the city, always leaving from the RSYC and walking along a narrow two-lane paved road that wound along the shoreline. Within minutes, we'd pass a small prison, where we looked curiously at any inmates that happened to be outside. After twenty minutes, we'd arrive at the central bus terminal; taxis and buses then filled the road, and often the warm air was heavy with the smell of diesel and exhaust fumes. Crossing through the terminal was an adventure all on its own, with dozens of buses coming and going. They didn't stop, didn't signal, and could be going in any direction.

After another ten minutes, we'd be downtown, where narrow streets curved along the shoreline, changing name frequently. Traffic moved so fast that I found myself running across the intersections. We would then branch inland, following streets that wound up and around low hills through older sections of the city and were chock-a-block with vehicles. Narrow storefronts, crammed tightly together, lined the sidewalks, while crowds of pedestrians flowed back and forth out front.

We felt very welcome, with native Fijians smiling and

saying hello when we passed on the street. A few even called out "Boolah!" from across the road. If we happened to pause uncertainly on a street corner or looked at a map, someone always stopped and gave us directions, whether we wanted them or not. I particularly noticed the women, as they were tall and well built, with thick, fuzzy hair standing out six inches from their scalps. I very much admired the confident manner with which they carried themselves.

On one of our first excursions, we located the offices of the Health Department and paid the $33 fee charged for clearing us in. Dave also checked out a few stores and found a 5-kg container of cabin crackers, for which he developed a life-long passion. One day we happened to be walking by a Cinema 6 theatre when it was going in; as movies changed frequently and prices were cheap, we became regular visitors.

We located the wet market in a huge building near the bus terminal; it provided all the fresh fruits and veggies we could possibly want. It was without doubt the most impressive such market we saw anywhere, with rows and rows of produce piled high on tables that stretched down the main floor. Wandering through the aisles, we had our pick of bok choy, cabbage, tomatoes, chili peppers, green beans, carrots, eggs, coconuts, bananas, oranges, papaya and pineapple.

Potatoes, onions, and other unusual items were sold in stalls on the upper floor, along with a huge variety of bulk spices. Many of the stalls also sold kava root, while a small kava bar occupied one corner. (We weren't too sure about this indigenous drink, which was reputed to have a sedating effect.) At the back of the building were two small shops that sold fresh fish, and a roofed area with concrete floors where the locals sat with their catch of the day (and hordes of flies).

On Fridays and Saturdays, the market expanded outside, taking over the sidewalks as well as a large open area down one side of the building. With tarps strung overhead, roots, leaves and other produce were piled high on pieces of cloth laid on the ground. Half the population of the city seemed to come to shop, and the sidewalk would be so crowded near the main entrance that it was difficult to find a way through. Fijian women looked neat and clean in colorful, flowered shirts and dresses, and Indian women stood out like exotic birds in beautifully tinted saris.

One day, we found our way to Cumming Street, which was only two blocks long, but the area always fascinated me. The shops catered to the Indian population and several sold beautifully fashioned jewelry, much of it 22-carat gold; some pieces were so finely crafted, they looked like spun gold. Other shops were draped with saris of varying styles and such a rainbow of colors that I couldn't help but stop and admire them.

The real attraction, however, were the crowds on the sidewalks, ebbing and flowing around stalls selling jewelry and watches. Walking, talking, and wearing all manner of dress, they provided a kaleidoscope of color and sound that was irresistible. I tried taking pictures but failed miserably, as I didn't seem able to separate myself from the hustle and bustle around me. I began to suspect that it was the flashing eyes and generous smiles of dark-skinned faces, the gleaming ripple of muscle in arm or leg as passersby dodged one another that kept my eyes dancing.

Small, slim Indian men with straight black hair slipped purposefully through the crowds, while their bushy-headed Fijian counterparts, clad in shirts and sulus, stopped and chatted, then ambled on. Most of the Indian women wore saris, and a few Fijian women in long skirts occasionally strolled past, while others wore western apparel, including fitted dresses and long pants. A few tourists, like me, sported shorts. Flitting around the edges were Fijian schoolgirls in uniforms that were probably intended to be shapeless, but actually highlighted their slim bodies and young breasts.

The Indian women were especially intriguing, some so small and finely boned as to appear fragile, their delicate features and dark skins accented by gold jewelry and the striking shades of their garments. Not all wore saris, as any style of dress seemed acceptable, but those who did either were wrapped up tightly or displayed a bare midriff. When I came across some beautifully tinted saris, I was tempted to buy one, but knew intuitively that I could never wear it with the natural grace of those around me.

In early July, a flurry of cruising boats arrived in port, so we refocused our attention on the RSYC. Amongst the new arrivals were Tony (*Dandelion Days*), Del and Joanne (*Limbo*), Ernie and Emily (*Quiet Time*), Bill and

Gail (*Bright Wing*), Steve and Molly (*Halo*), Charlie and Janette (*Quark*), and Ed and Fran (*Aka*).

It was now impossible to walk through the clubroom without being caught up in discussions about equipment, cruising areas, or weather. It turned out that Ed had the expertise needed to fix our depth gauge, while we had a spare manual water pump that was just what Tony needed. Although I didn't recognize him, Charlie was the mechanic from the outboard engine shop in Opua, who showed me how to disconnect the kill switch.

A few families also showed up in the clubroom, and I was finally able to satisfy my curiosity about kids on sailboats. They all seemed healthy and reasonably well behaved, and one day, I watched in admiration as several youngsters took turns shinnying three or four feet up a veranda post. It was apparent that life was life, however you lived it, and these families just happened to live it on a boat.

One boat, *Three Daughters*, had four children onboard, the youngest being a little boy, maybe four years old. He frequently seemed to try his dad's patience because he couldn't stay away from a cat that hung out near the clubhouse. In all fairness, I think the cat was as much to blame as the boy because I never saw it except when the young lad came ashore, and then it was always front and center.

The weather changed during the second week of July, when a mass of warm, humid air moved over the island. Sudden squalls brought clouds, rain showers, and gusty winds that caused a few boats to drag their anchors. Strong southwest winds also pushed swells up into the harbor, which caused the boats to roll. I started paying closer attention to boats anchoring nearby, which proved effective, as only once did we have to ask a boat to re-anchor.

On our eighteenth day in Suva, the weather fax finally arrived. We rushed off to the Fed-Ex Customs Depot downtown, but then spent half the day caught up in the tedious procedures imposed on overseas boats by local authorities (probably because we were exempt from taxes). The depot first sent us over to the main Customs office to obtain an authorization form. When we returned half an hour later, with form in hand, a clerk handed over a large box, with instructions to take it to the Custom's Boarding Officer at King's Wharf.

The box wasn't all that heavy but was very awkward to carry, as it was at least six times the size of the fax machine inside. Despite the fact that the day was very hot, Dave then balanced it on his shoulder, and we paraded through the streets and down to the wharf. The boarding officer checked our documents, ensuring the shipment was actually going onto an offshore boat. He then signed the form, leaving us free to walk the mile back to the RSYC.

The next day, Dave secured the unit to the chart table and attached a splitter to the HF antenna, so we could direct the signal to either the radio or the fax machine. I studied the operating manual, including the types of weather maps available and the times of transmission. From then on, twice a day, I listened to voice transmissions and compared the information I heard to what I saw on weather maps produced by the fax machine; gradually, I started to understand what they meant.

Time now started to drag but we were still waiting for the Hydrovane order to arrive, so once again we turned to our "to do" list. Dave removed the broken car from the whisker pole track and had it spot-welded, then tore the head apart, as it had developed a slow leak. With 80% humidity in New Zealand, I'd noticed mold growing on a few surfaces before we left, so started scrubbing lockers. When we got up to a dry sky one morning, we worked on the cabin windows, cleaning them and running a bead of silicon around each one, both inside and out.

The anchorage was actually very dirty, with smoke and exhaust residue settling on the boat each night. I was then scrubbing down deck and cockpit once a week, re-using the rinse water after washing clothes. Diesel spills were common, too, with large pools of streaky water sending fumes wafting into the cabin as they drifted about with the tide. One day, we got up to find an extra layer of black grit on deck, so I started scrubbing even before breakfast. Later, we learned that a freighter had offloaded a cargo of coal overnight.

After sitting through a blustery storm another morning, we set off for the Suva Museum, which was an hour's hike across town. We found the maritime history of the islands to be of particular interest, as some of their sailing canoes had been immense. However, just how they managed to steer them remained something of a mystery.

The exhibits dealing with cannibalism initially

appalled me, but something about the matter-of-fact style of presentation managed to pique my curiosity. I then read about "human meals" and how they were prepared, and studied the "cannibal forks" used to eat the brain. We also learned about life in traditional villages, where men started each day by attending a *yoqona ceremony*, at which a priest drank kava and articulated messages from the gods. Men's bodies had carried many scars, most of them self-inflicted, while women's bodies had been heavily tattooed from waist to mid-thigh and around the mouth.

We were most shocked, however, by the recent political history of Fiji, as a coup d'état had occurred in 1987, just ten years earlier. The prime minister and his entire cabinet were kidnapped during the opening session of parliament, just one month after a general election. That such a thing had occurred in a country with a history of democratic governance similar to our own was very disturbing. But then, so was the fact that neither of us knew anything about it, although we considered ourselves to be well informed.

The unrest, we learned, was rooted in racial tension dating back over a century, to a time when the British brought laborers from India to work in the cane fields. Policies then in place allowed only native Fijians to own land, and protected them from forced labor. Over the years, however, the Indo-Fijian community had increased in numbers, wealth and influence, and it was their success at the ballot box in the 1987 election that resulted in the coup. A new constitution was then adopted in 1990, one that highly favored native Fijians and mandated that the prime minister be one. So, thousands of Indo-Fijians left the country, although we were told exchange controls limited their ability to take their wealth with them.

As it happened, legislators were debating a constitutional amendment while we were in Suva in 1997, one that was supposed to create more equality between the races. One day, we read in a newspaper that the bill had passed; politicians were very pleased with themselves and busy congratulating each other. There was even talk of reapplying for membership in the British Commonwealth, as the country had been forced to withdraw after the coup.

Newspaper editorials weren't as reassuring, however. At least one claimed the amendment would not change anything, that clauses appearing to address specific issues were overruled by clauses maintaining the status quo. Some residents bitterly claimed the whole process was too confusing and bureaucratic, that leaders were only interested in their own futures.

When we returned to Suva two years later, in August of 1999, continued political uncertainty had noticeably disrupted the tourism industry. Two more coups rocked the country before the end of 2006, and it wasn't until 2014 that the situation apparently began to stabilize. Reports indicate that the economic cost of the conflict was immense, especially as many of the best and brightest minds had emigrated.

CHAPTER 21

Rain and More Rain

Low, overcast skies settled over Suva in the latter part of July, bringing grey days of constant rain that made life miserable for those of us living in the harbor. We found ourselves deferring trips ashore as the ride could be miserably wet. But the return trip was often worse, and one afternoon we sat in the clubroom for two hours, waiting out a deluge. Sudden squalls caused boats to drag their anchors, and occasionally, one would run afoul of a neighbor's anchor chain. Strong southwest winds also pushed swells in through the reef, and one night *Windy Lady* rocked and rolled for hours.

All the rain meant we collected buckets of water off the dodger roof, but also had to bail out the dinghy two and three times a day. Although I tried to choose the best days to do laundry, clothes never dried outside, so towels and sheets would then be hanging off the grab rails inside the cabin. We were now baking bread and cookies onboard too, using up some of the supplies we'd brought from Canada.

However, we still had to make trips ashore to visit the wet market and buy meat, although we were buying less of it, as we didn't really care for the beef available in Fiji. We also hauled fresh water from the RSYC to replenish our water tank. After filling it when we arrived, we were able to keep it topped up with just one trip each week, using two 22-liter jugs.

Dave managed to keep busy doing chores, whether it was cleaning the bilge, checking batteries, changing engine oil, or such like, but I was getting bored. I started researching our next ports of call and remembered that we would be traveling in areas where malaria was endemic. Put off by the side effects reported by some cruisers, we now decided not to take anti-malarial drugs. That meant we needed an alternate line of defense, so I started thinking about mosquito netting for the companionway and two hatches.

On our next trip downtown, I badgered Dave into locating a dry goods store, where I bought yards of netting. While the project seemed simple enough, nothing ever was onboard *Windy Lady*. Curtains to cover the companionway had to be seven feet long and full enough to seal around the edges of the doorway. The hatch coverings had to fit snugly around the deck openings too, so nothing could crawl through. By trial and error, with a bit of rope, bungee cord, and help from Dave, we ended up with billowy hatch coverings and long, loose curtains over the companionway that served us well for years.

Two packets of mail postmarked June 30 eventually arrived, but neither were the one we expected. When Dave phoned our mail forwarding service in Canada, they assured him the missing package was sent on June 13. When he dug deeper, he learned that it went surface, which meant it would take at least two months to reach us.

As the days dragged on, we grew more and more tired of sitting around in the rain, waiting for deliveries that never came. We knew there were worse places to be, as the weather was bad everywhere, but still we got up one morning, looked out at the rain, and decided to leave on the next sunny day.

First, we found our way to the Australian consulate and obtained our visas; we'd been told it was easier to obtain them in Suva, rather than in Vanuatu or New Caledonia. We also purchased a cruising permit that covered most of Fiji, as we hadn't been able to decide just where we would sail. On August 1, we cleared with Customs and prepared the boat for sea. Two days later, while still waiting for a glimpse of the sun, the outboard engine died and Dave couldn't get it running. Rowing ashore, he loaded it into the trunk of a taxi and took it to a dealer on the far side of town. Another two days went by.

On August 6, we ran into Ken and Pat on *Iron Butterfly*; they had just arrived from Tonga. We then stood on the sidewalk between rain showers and learned of their cruising adventures that season. After leaving New Zealand, adverse winds kept them from reaching Niue, Cyclone Kelly then threatened them in Tonga, and rough seas had made the crossing to Fiji unpleasant.

The next day, a storm brought 40-kt winds tearing through the anchorage and most yachtees stayed onboard. When we talked to Ken the following day, he had devastating news. A cruising boat called *Camelot* had been lost the previous morning; it had gone up on a reef near the entrance to Blight Water, on the north side of the island. The report had a chilling effect on us all, a reminder of just how quickly life could change.

Over the next few days, we pieced the story together. The mishap occurred about midmorning, with the boat making seven kt and the autopilot doing the steering. Initially, there was no reason for concern, as the sea was calm and the sky overcast. As the tide was then ebbing, the captain chose to wait until it turned before trying to refloat the boat. But one of those fierce, fast-moving storms that we'd been experiencing then blew through. Exposed up on the reef, the boat started to break up half an hour later. The crew, carrying little more than their passports, was taken aboard another cruising boat, *Silver Cloud*.

Later that day, we sat for an hour in the clubroom, which was abuzz with the news. Before we left, I checked for messages at the office and found one from the marine store in Victoria. Although we had heard nothing since the original phone call six weeks earlier, they had shipped an alternator three days before. When Dave checked the attached invoice and saw that shipping charges of $336 had been added, his roar could be heard across the anchorage. He first wanted to send it back, but after sober second thought realized he'd have to pay for the return freight, too. Besides, he kind of wanted it.

Four days later, the alternator arrived, and we followed the same tedious procedures as before. This time, however, a guard challenged us as we entered through the main gate at King's Wharf, asking to see our paperwork. We then continued on to the office, where the Boarding Officer signed the documents. As another gate was closer to the RSYC, we left through it, waving at the guard as we went by.

Two blocks down the road, a vehicle came roaring up from behind and cut across the sidewalk in front of us. Two uniformed Customs officers jumped out, demanding to see our parcel and paperwork, then checked it all very carefully. After exchanging a few words, they shrugged their shoulders and handed the stuff back, then returned to their vehicle and drove away. Feeling somewhat bemused, we could only assume that procedures had changed since our earlier visit. We probably should have checked with the guard when we left.

Next morning, sunshine was poking through a few holes in the clouds. It was then August 13 and we'd been in Suva seven weeks. It was time to go. We quickly dinghied ashore, disposing of garbage and filling our water jugs, then winched the dinghy aboard— but couldn't free the anchor. Without even thinking about it, we just used the weight of the boat to break free.

We motored out through the passage in the reef shortly after 1000, heading for Beqa Lagoon, some 37 nm south and west of Suva. Earlier, we had purchased a cruising guide to supplement our charts, and now, with the forecast calling for unsettled weather, we decided to limit our sailing to the island of Viti Levu. We would spend three days sailing along the south coast, then pass through the barrier reef at Navula Passage, and continue cruising in protected waters on the west side.

With blue skies overhead and winds behind us, I was looking forward to a pleasant, sunny day on the water. That wasn't to be however, as high, rough seas bounced the boat around viciously for hours. Surf was pounding on the reef as we approached Sulphur Pass, but with help from the cruising guide and GPS, we safely entered the lagoon and made our way down to the anchorage on the west side of Yanuca Island.

By 1600, *Windy Lady* had joined two other sailboats and was floating quietly in calm waters in the lee of the island. Glad to be out of the city, we sat in the cockpit, enjoying the peace and quiet of our surroundings. After sunset, the evening was extraordinary, with bright stars filling the sky and a warm wind caressing our faces. Before morning, however, swell crept in around the end of the island and the boat started to roll.

We were up early and underway shortly after 0730, following a deep-water channel through Yanuca Passage. The day was perfect, with clear, sunny skies, but winds were light. Staying about two nm off the shore, we motored west and managed to get decent sunburns as we relaxed in the cockpit. Clouds then moved in during the afternoon, and after traveling 44 nm, we anchored in Natadola Harbor just before 1600.

I was not happy with the anchorage right from the start, as it had a wide entrance opening to the southwest. If winds happened to shift, swells would push right up to the top. But clouds were growing heavier and darkness not far away, so we really didn't have a choice. Rain started to fall just after we went to bed, and the first swells reached us about midnight, rolling the boat slowly from side to side. As the swells grew higher, the boat rolled farther, and we slept less.

The early forecast next morning called for winds of 30 kt by midday. Anxious to get underway as soon as possible, we were out on deck at first light, checking on conditions. But I looked around with sinking heart, as low, grey clouds and rain completely enveloped the boat, obscuring everything around us.

When the shoreline began to emerge from the gloom an hour later, we started the engine and raised the anchor. While still in the channel, *Windy Lady* was hit by 30-kt winds on the beam, heeling her over, and then large swells tossed her about roughly when we reached open water. As we made the ninety-degree turn to the west, however, she rolled over so far that we were grabbing onto whatever was handy. Once on route, with wind and waves behind her, she settled down quickly.

Staying about one nm offshore, we motored westward, navigating through a grey, shadowy world, where clouds lifted and re-settled on the hillsides, and passing rain showers reduced visibility to zero. Two hours later, as we approached Navula Passage, the sky started to lighten.

Using binoculars, I easily spotted the marker buoys on the edge of the reef.

The sea calmed quickly after we turned into the passage, although winds held at 25 kt. Once through the barrier reef, we followed range markers into an all-weather anchorage and dropped the hook at 1235, finding good holding in eighteen feet of water. After a sleepless night and anxious three-hour passage, I was just relieved to have the boat in safe harbor.

Initially, we shared the large anchorage only with the pilot boat stationed there, but late that afternoon, another cruising boat crept in, looking for shelter. After a good night's sleep, we enjoyed a leisurely breakfast in the cockpit next morning and discussed our next move. We were then in sheltered waters behind the reef, in the lee of the island, so there seemed no reason not to continue to Musket Cove, which was about three hours away.

We were a tad optimistic, however, as under an overcast sky, the dark waters continued to hide their secrets. Somewhat hesitantly, we motored around rocky shoals and small islands and, even with the help of the cruising guide, missed the channel into the resort. After plotting our position on a chart, we backtracked and identified Black Rocks, on the north side of the entrance. We then followed marker buoys down a long, twisting channel that, on the surface, looked no different from surrounding waters.

All went well until the resort came into sight off to starboard. Dave, who was at the helm, then started steering towards the boats that he could see anchored in a small cove. I was on lookout duty at the bow and repeatedly warned him to keep to port, but the lure of the boats was too strong. When he turned the helm towards them one last time, the depth gauge shot up twenty feet (from sixty to forty feet).

He corrected his course immediately, but by then it was too late. While the stern may have been in forty feet of water, the bow was over the top of the reef. I could clearly see the coral beneath my feet, and in my head, could hear the keel grinding harshly upon it. Luck, however, was still with the captain; there was just enough water over the reef to keep *Windy Lady* afloat.

Close to forty boats were squeezed into the anchorage, but Dave managed to find a hole on one side. After dropping the hook in fifty-five feet of water, we couldn't release as much chain as we'd have liked because of nearby boats. After waiting a bit for *Windy Lady* to settle, we put the dinghy in the water and went ashore.

We saw only a few people as we walked about the resort, but on the beach ran into our friends off *Limbo*. After discussing the reefs we'd seen on the way in, or rather hadn't seen, Del admitted, "Yeah, we went up on a reef this morning. But it was nothing serious as another boat was nearby and was able to pull us off."

The afternoon had turned hot and muggy by the time we returned to *Windy Lady*, so we opened the hatches wide, trying to catch a breeze. By bedtime, dark clouds had moved in and the barometer was dropping, but the temperature inside the cabin was barely tolerable. Although we talked about closing the hatches, we decided against it, as Dave pointed out, "We're a boat, aren't we? A little water won't hurt us!"

He had cause to rethink that statement about midnight, when we woke to the sound of rain drumming on the deck. Crawling out of the berth, he went forward to close the two hatches. Coming close behind, I could see the teak floor shining wetly in the dim light as he reached up to close the galley hatch, then heard him swear softly as he stepped into ankle-deep water in the head. After closing that hatch, he groped about on the floor and removed the plug to the holding tank, letting the water drain in. Meanwhile, I found a basin and started mopping up in the galley; together, we wiped up four full basins.

Come morning, we wiped down the floors again, as well as the surrounding walls, but it took two full days in a hot sun before the seat cushions dried out. Of course, the dinghy was also full of water, with seats awash and low waves threatening to run over the transom. The weather broadcast that morning reported eleven inches of rain had fallen overnight.

When the clouds cleared away, the winds picked up, sweeping through the anchorage at 20–25 kt for three of the next four days. Concerned that we might be dragging the anchor, I spent a couple of hours in the cockpit, tracking our position with the GPS. While I took readings, Dave sat and studied the cove; the tide was going out, exposing more and more shoals. Finally, he shook his head, wondering aloud, "How did we ever manage to find our way in through all that!"

When we went ashore the next day, Dave bought

a membership in the Musket Cove Yacht Club. He'd wanted to do so ever since he read that member names and boat names were carved into the beams of the clubhouse. Memberships were actually cheap, the rational being that it cost plenty to get there. After spending some time on the beach, we had a very wet ride back to *Windy Lady*, with winds gusting up to 30 kt. A few boats had moved on, leaving a little more room, so Dave let out another fifty feet of chain.

The next day brought a respite from the wind. With the promise of clear skies and warm sunshine, I was soon digging out our snorkeling gear. But we had disposed of the inflatable in New Zealand, so first had to add an extension to the boarding ladder. Once we were able to climb out of the water onto *Windy Lady*, Dave then spent two hours cleaning the transducers for the Signets. By the time we swam across to the reef, the sun was high and so was the tide. With fine sand suspended in the water, visibility was poor; even more disappointing, much of the coral was dead.

The winds returned the following day, and by noon, half the boats in the anchorage were gone. With more wind in the forecast, we saw no reason to stay either and left the next morning. With the sun high and bright, it was then easy to see the shallow water over the reefs, so easy to stay in the narrow, winding passage between.

We raised the sails once we were clear of the channel, but half-an-hour later lost the wind in the lee of the main island. We then motored for four hours, leaving the steering to the autopilot, while we soaked up warm sunshine and enjoyed views of tropical islands and calm blue waters. The wind picked up just as we entered the harbor at Lautoka, making it difficult to anchor. With a mud flat extending well out from the dock, it was going to be a long dinghy ride ashore.

Next morning, *Windy Lady* was rolling in swells pushed up by a 20-kt wind. When we went up on deck to let out another fifty feet of chain, I just stopped and stared in disbelief. The entire top of the boat was covered with a layer of black soot. So while Dave went forward and released more chain, I started my day by scrubbing the cockpit. We later learned that it came from the sugar mill.

We dinghied ashore just before noon and checked out a nearby marina, which was rumored to be in receivership. The facilities were in tough shape but at least the toilets worked, and Dave was able to locate a telephone. He called the yacht club in Suva and learned that our missing mail had arrived, so instructed them to forward it. A couple of days later, we picked it up downtown, but there was nothing in it to warrant all the fuss.

We spent five days exploring Lautoka, wandering up and down the streets for hours. We also passed through a small village on the way in, but never saw a big smile or heard a cheerful, "Boolah! In fact, the atmosphere in the city was very different from what we'd grown accustomed to in Suva. At first, I thought the city was just a modern industrial town; many people worked in the sugar mill, and stores were newer, larger, and less personal. But everybody on the sidewalks seemed preoccupied, and after a few days, we began to wonder if racial tensions might be a little closer to the surface in that part of Fiji.

With continuing strong winds, the dinghy ride ashore was frequently wet, and we broke another shear pin at low tide when the prop hit something in the murky water. The winds also brought another ten boats into harbor, bringing the total to twenty-three. We visited with some of the newcomers on the deck at the marina, learning that stormy conditions had limited their cruising and kept the sand stirred up over the reefs for most of the season. Wandering through the adjacent hardstand, we also found people working on a few boats. Nearby, several other boats gave new meaning to the term "hurricane hole". They were sitting in holes in the ground and, with keels hidden from view, appeared to rest on the grass.

We had hoped to visit the Yasawa Islands on the far west before leaving Fiji, but with wind and rain, there didn't seem to be much point. As we were running out of things to do, and cruising boats were starting to depart for Vanuatu, we decided to move on also. We cleared with Customs and Immigration on August 26, and departed the following day.

CHAPTER 22

Passage to Vanuatu

In the early morning of August 27, we were up on deck, raising the anchor. With a clear, bright sky, it looked like a good day to start a passage. As winds were light in the lee of the island, we then motored for four hours as we made our way through quiet waters to the western passage through the barrier reef. With the autopilot doing the steering, we relaxed in the cockpit, enjoying the warm sunshine and each other's company while we had the chance. We now knew that such occasions were rare at sea.

Still, we kept to our usual duty schedule, and I was on watch when *Windy Lady* suddenly veered off course. I jumped up and took control of the helm, while Dave went below to check the autopilot. It had malfunctioned a time or two before, but he'd found no obvious reason for the problem. We referred to it as going walkabout, as it seemed to steer the boat around in a large circle. Now he fiddled with the controls for a bit and eventually it started working again.

The sky was still bright and sunny when we stopped on the lagoon side of Malolo Passage and raised the mainsail. We put in a reef too, as we'd concluded that the full sail was too much for Otto to handle. For some reason, we believed that winds would be light beyond the reef, so set the staysail. We expected to be at sea for five days, as the distance from Lautoka to Port Vila in Vanuatu was 542 nm. We would be heading due west, with SE trade winds on or abaft the beam, and anticipated no difficulties.

Shortly after midday, Dave took the helm and steered the boat out through the passage in the reef. With the bow cutting smoothly through waters ruffled only by a light breeze, I sat back, enjoying the views of breaking surf. As we exited the channel, I guess we left the lee of the island because suddenly all hell broke loose. South winds gusting from 15–25 kt began pummelling the sails, while turbulent five-foot seas tossed the boat all over the place.

Caught completely by surprise, Dave struggled to control the helm, while I just tried to hang on. Once we were clear of the entrance, he turned the bow on route and the boat settled down somewhat. As usual, I took the helm and pulled out the headsail while he went below to shut down the engine. But as the wind filled the sails, the boat speed picked up, and soon *Windy Lady* was charging through the humps and bumps in front of us. There was no way I could settle her down, so when Dave reappeared, I suggested that we take down the staysail. He chose to fight with the helm for another hour before deciding to do so.

Otto still couldn't handle the rough conditions when

I took the helm at 1600, so I put a reef in the headsail, but that didn't help. Meanwhile, Dave had gone below to fill in the journey log. Moments later, he poked his head out into the cockpit and calmly announced, "The GPS isn't working and I don't know what's wrong with it."

I just stared at him in disbelief, and then icily responded, "If that's the case, we should turn around, go back to Lautoka, and have it repaired. I don't want to start on a voyage with a sick GPS!"

He turned away without answering, leaving me standing at the helm, scowling and thinking very black thoughts. We followed the GPS blindly at sea, so I was actually quite angry. How would we ever find a small island in that vast ocean? I was also concerned that he might think my hit-and-miss attempts with the sextant might actually be of some help.

Half an hour later, he poked his head outside again and said with forced cheerfulness, "The GPS is working now, so everything's going to be find." Of course, he knew that I knew that in spite of whatever he'd been doing, he had no idea why it had quit, or why it had started working again.

Winds gusted from 15–25 kt all that night, and sea conditions deteriorated as the hours past. Initially, we were able to sit and assist the vane, but by morning, we were standing at the helm, hand steering. So much spray was coming into the cockpit that we wore foul-weather gear on watch, and boat movements were rough and erratic, making it impossible to sleep. I probably wouldn't have anyway, as I was much too worried about the GPS, positive that it would again quit working.

Shortly after the start of my 0800-watch, winds began gusting up to 30 kt. I was then too busy steering the boat to worry about the GPS. All that day, *Windy Lady* charged across the ocean at seven kt, rolling, pitching and twisting, sometimes quite viciously. Again, it was imperative to hang onto the boat, even if moving only a few feet. On the positive side, however, the salon windows weren't leaking. At noon, after our first twenty-four hours (Day 1), we logged 150 nm.

Knowing I wouldn't be able to rest, I spent that afternoon in the cockpit, sitting with my legs braced to stay in one place. Dave stood at the helm and hand-steered, but constantly scowled in the direction of a sailboat about a half mile to port. That boat was *Limbo*, with Del and Joanne aboard. They'd come through the reef an hour behind us and slowly been gaining ground ever since.

I couldn't believe how much the two boats were moving around in relation to one another. Sometimes *Limbo* was off our bow, sometimes it was abaft the beam on the port side, or it might be anywhere in between. With extremely rough seas over six feet high, this was just a visual indication of how much *Windy Lady* was turning and twisting. In spite of Dave's competitive instincts, he now agreed to put a second reef in the main, so I brought the boat to a stop while he worked on the sail.

This was our introduction to reinforced trade winds, a term we'd first heard only a few weeks earlier. Winds gusted from 15–25 kt, even up to 30 kt for much of the passage, while backing from south to SSE to SE, which put them on or abaft the beam. A bigger problem for us, however, was the SE swell, which surged relentlessly against the port quarter and grew to heights we'd never before seen. Because we were heading due west, wind and sea conspired to turn the bow southward, overwhelming Otto, so that we hand-steered for much of the passage.

By nightfall, winds had backed to the SSE and dropped to 25 kt, but seas continued to build. When Dave relieved me at 2000, seas were over eight feet, and I'd put a second reef in the headsail. Still, I wasn't able to sleep between then and midnight, as the berth constantly jerked beneath me. Not sleeping wasn't all that unusual, but the constant twisting and jolting of the boat left me feeling achy all over, which added to the fatigue.

I had to force myself from my berth at midnight, feeling much more tired than usual. I guess Dave felt much the same, because when I stepped into the cockpit, he immediately went below, pausing only long enough to say, "The GPS quit working two hours ago."

Standing at the helm, I stared out into the darkness, trying to hold the bow on course, and bleakly wondered how we would ever find our way into Port Vila. Now that I was over my anger and actually thought about it though, the answer was quite simple. Before long, I had even convinced myself that it wouldn't be that difficult and was looking forward to the challenge.

Thanks to Dave's decision to maintain a journey log, we had a starting point; our last recorded position had been two hours before the GPS quit working. Since then, we had steered a course of 270 degrees, and the Signet log had recorded the distance the boat travelled through the water. We also had pilot charts that showed the speed and direction of currents in the area. The only real unknown was the effect of the swell pushing against the port quarter, and I figured that a minor problem. As my confidence returned, I felt quite cheerful, telling myself that if we happened to miss Vanuatu, we were bound to run into Australia.

Dave, meanwhile, was lying awake in his berth; he was also thinking about the GPS, but his thoughts were a little more practical. He'd performed a self-test earlier and it indicated an internal problem, but when he entered a new waypoint, the unit recalculated course and distance. That made him think the problem was with the antenna. The longer he thought about it, the more convinced he became that it just wasn't picking up satellite signals.

At first light, he moved a few feet away from his post at the helm and took a closer look at the small, mushroom antenna clamped to the push pit rail. Wondering whether it could have moved, he released the clamp, repositioned the unit, and re-secured it. When he checked the monitor a few minutes later, the screen was lit up brightly, showing boat position, as well as distance and course to destination.

When I came on watch at 0800, there was good news all around. The GPS was not only working, but for the first time, winds were down to 20 kt from the SSE. Otto was even making an effort to steer the boat, although I still needed to provide constant assistance. As the sky was sunny, I decided to make a serious attempt at taking a sun shot at noon, and see how close I could come to our GPS position. I then asked Dave if he'd come on watch a half hour early.

By mid-morning, however, seas had grown to eleven feet, and I was again standing at the helm, hand steering. With the boat bouncing about in rough seas, I wondered whether it was even worthwhile bringing up the sextant, but when Dave showed up as agreed, I felt compelled to try. Hanging onto the grab rails, I then went below and unpacked the instrument.

As I needed both hands to manipulate it, the first challenge was finding a place to stand where I wouldn't have to hang on. Looking around the heaving deck, the only possible spot was along the centerline, just forward of the binnacle. By leaning back against it and bracing my legs, I could absorb the movements of the deck with my knees, which gave me a fairly stable base. I then raised the sextant to my eye, adjusted the various settings and, using the mirrors, brought the sun down to the horizon.

The first readings varied far more than normal, so I took a few minutes to try to figure out why. It was then I noticed that the rolling motion of high swells in the distance actually caused the horizon to move up and down. I took a few more readings, trying to compensate for that problem, when the sun went behind a cloud. Stopping to change the lens, I had just raised the sextant again, when a wave reared up behind me and dumped half-a-bucket of saltwater over my head. I managed to keep the sextant dry but my time had run out; the sun was now approaching the zenith. All I proved that day was that even though it was sunny, it wasn't possible to take a sun shot.

For Day 2, we recorded 141 nm. After plotting our noon position, I checked the numbers for the ten-hour period we'd been without the GPS. According to the Signet log, the boat had traveled 60 nm through the water, but the GPS revealed we'd covered 66 nm over the ground, so the currents had been strong. We had also drifted one nm south of our course line, which wasn't surprising. Still, even with two days to go, I figured we would have found our way in to port.

Early that afternoon, the winds again picked up, gusting from a low of 15–18 kt to a high of 28–32. Extended gusts lasted up to ten minutes, making it even harder to stay on course. When I relieved Dave at 1600, we estimated seas at ten to twelve feet. After standing at the helm for a few minutes, with swells rolling up behind me, I decided to try steering with my back to the bow. That way I could look out over the port quarter and study the line of swells rolling out of the southeast.

These weren't long swells, as I'd seen on the way into New Zealand, but small mountains of water that loomed high over the boat. They then surged beneath the hull, causing the deck to roll and pitch beneath my feet. I watched, mesmerized, as the swells grew higher and higher, until I actually felt intimidated. I tried to estimate their size by comparing them to the dodger roof and radar mast, and my best guess was twenty feet. After about half-an-hour, these huge swells gradually began to subside. By the end of my watch at 2000, the seas were back down to twelve feet, but I'd put two more reefs in the headsail (for a total of four).

Wind speeds remained unchanged through the dark night hours, but they backed to the southeast about midnight, and seas began to ease. By 0400, the swell was down to eight feet, and soon Otto was making efforts to steer the boat. When I returned to the cockpit at 0800, the swells were five feet, and Dave had shaken three reefs out of the headsail. At noon, the breeze was steady from the SE at 15–20 kt; we logged 127 nm on Day 3.

The winds remained at 15–20 kt, occasionally gusting up to 25, during the last twenty-four hours into Port Vila. Seas were up over six feet for about eight hours during the night, so we again hand-steered. When Dave relieved me at 0400, we were right on course and expected to be in sight of land by daybreak. As we wanted two sets of eyes going into the harbor, I stayed in the cockpit.

As the sky lightened, I searched in vain for land, but saw only waves stretching to the horizon. My first thought was that the GPS had failed again, which wouldn't have surprised me in the least. A few hours later, the southeast corner of Efaté Island came into view; it was low and flat, with only the treetops showing above the horizon. As usual, we motored the last few hours into port, charging the batteries.

Dave contacted Port Vila Radio about 1000, giving them an ETA of 1200. Almost immediately, *Limbo* came up, advising they had just anchored, while two other boats, *Idiom* and *Windseeker*, arrived in port just ahead of us; we had all left from Fiji. We dropped anchor a fair distance out, as cruising boats were crowded close together near the dinghy dock.

Even though it was Sunday, the authorities were busy clearing in new arrivals. While we waited, we flaked the mainsail, and I wiped up the salt spray that coated the cockpit. Customs eventually gave us 30-day visas, while Agriculture took the few vegetables we had left, leaving only a couple of onions. After the officials were finished, we tidied up the boat, putting away sheets and equipment.

We proceeded to toast our safe arrival, both of us relieved that the passage was over. Strong, gusty winds and lumpy seas had tied us to the helm for most of the crossing, and prevented us from resting when off watch. Although the first three days had been warm and sunny, we'd worn our foul-weather jackets throughout because of the amount of spray thrown into the cockpit. At night, we'd pulled on long pants and I'd worn a sweater, too, as the temperature dropped after sunset.

We arrived in Port Vila at noon on August 31, having spent 100 hours at sea; we logged 564 nm, for an average speed of 5.6 kt. On our way into harbour, we noticed inconsistencies between the chart and the GPS. When I then plotted our GPS position on the chart, it showed us to be on land. When we checked further, we saw that the chart was in error by almost two nm in both latitude and longitude.

After supper that night, Dave fiddled with his transistor radio while I relaxed with a book. He was able to tune in only one station and it was in French. As he played with the dial, we caught the words "Princess Diana" and "morté". We couldn't understand anything else, but heard those words repeatedly. Obviously, something significant had happened.

CHAPTER 23

Port Vila and Efaté Island

After making our way ashore next morning, we checked out the cafes, restaurants, and bars that lined the main street in downtown Port Vila. Although it was very quiet at that time of the day, every TV in every facility was tuned to CNN, and native staff on duty all stood nearby and watched. Stopping at the Office Bar, with its three large-screen TV's, we then learned of the car crash in Paris that had taken the life of Diana, the Princess of Wales.

The coverage continued for days and Ni-Vanuatu, as local people were called, spent their lunch hours watching CNN, seemingly mesmerized by a story that was repeated over and over. I found it hard to understand why they were so absorbed by the death of a woman they really couldn't have known much about. We then learned that people in cities around the world responded in much the same way. Only then did I begin to appreciate the power that the international media had to capture hearts and minds. I found it disturbing because the islanders had only gained their independence in 1980, and in a way, they seemed to have been colonized again.

The first Europeans to arrive in the islands had been the Portuguese; they landed in 1606 but didn't stay. De Bougainville then claimed the northern islands for France in 1768, while Captain Cook claimed the southern ones for Britain in 1774. He named them the New Hebrides. Each of these two countries then ruled in its own language, following its own customs, with missionary schools educating students in one of the two languages. In 1906, the two countries agreed to manage the islands jointly through the British-French Condominium, which formalized the duplication of all government services, except for a joint court.

With 113 native dialects spoken on 82 small islands spread out over 700 nm of open water, many of the islanders still couldn't speak to their neighbors. However, during the late 1800's, a pidgin language developed amongst laborers from Vanuatu and other Melanesia countries, who were working in sugarcane plantations in Australia and Fiji. Known as Bislama, it grew to become the second language of the majority of Ni-Vanuatu. Hence, when the country became independent in 1980, it recognized three official languages, French, English, and Bislama.

Because we came from Canada, with its own French-English tug-of-war, and had lived through a divisive national referendum just five years earlier, Dave and I immediately empathized with the islanders. We couldn't even imagine the bureaucratic nonsense that must have existed under a Condominium administration (pandemonium, we heard it described). As a result, I think our interest in the country differed somewhat from other cruisers, many of whom focused on visiting dive sites and traditional villages, or reliving the events of WW II.

Of course, we weren't aware of this initially and spent our first week in Port Vila doing touristy things, while recovering from the demands of the crossing. As we explored the city, flowers were in bloom everywhere, particularly bougainvillea, hibiscus and frangipani (plumeria), and the fragrance of the latter followed us up and down the streets.

We soon located a supermarket, then the wet market, which was set up in a large, open-sided, roofed building. Piles of coconuts, roots, and leaves were set out on long tables, but vegetables weren't very plentiful at the time. Several varieties of lap-lap, food wrapped in leaves and cooked in an underground oven, were also for sale.

In general, Ni-Vanuatu were slimmer and shorter than Fijians, with the tight woolly hair and broad noses of SE Asia (Melanesian). Many of the women wore knee-length, shift-style dresses with short sleeves, lace trim on the front, and foot-long ribbons flying off the shoulders. Some people wore flip-flops, others were barefoot; some met your eyes and smiled, others simply walked blindly past.

Women were far more visible in government offices than had been the case in either Fiji or Tonga. They didn't just have supporting jobs either, as one issued our inter-island cruising permit. We then ran into a Kiwi running a charter boat operation, and he explained that only Ni-Vanuatu could work in the country. Non-native people had to apply for permits, which were approved only if local people couldn't do the work.

Vanuatu had been a tax haven since the early 1970's, and the financial center in Port Vila appeared to be located in the middle of the tourist area. At least, I noticed a number of small offices, with women working at computer terminals, located next to souvenir shops, restaurants and travel agencies. At the same time, along the main thoroughfare, villagers sat on the ground with trinkets and shells displayed on dirty sheets of cloth in front of them.

Even as we explored the city, *Windy Lady* remained our first priority. The crossing had taken its toll, and we inspected her from top to bottom the day after we arrived in port. Our first task was to haul fresh water from shore and wash the salt from our foul weather gear, as well as from the deck and cockpit. Dave next replaced the Racor fuel filter, as the rough seas had again dislodged black gunk from the bottoms of the fuel tanks.

The boat had also developed a leak, causing the bilge pump to come on every eight hours. Dave traced it back to the rudderstock, which came through the hull beneath the port berth. After we'd dragged the mattress out into the main cabin, he removed the cover over the access port and found that both the inner nut and locking nut were loose. That had us digging into other lockers, looking for the large wrenches.

As soon as we found them, he knew that he had another problem; the tools all had long handles and the access port was barely a foot square. He set to work anyway, and soon a steady stream of loud, ill-tempered cursing had turned the air blue. Knowing I couldn't help, I hid out in the galley. His frustrated ranting only stopped after he'd made the last turn on the lock nut.

We soon recognized that we'd anchored near a traffic lane and every night, just after dark, a large vessel squeezed between us and two other sailboats. Without a light in the cockpit, we felt a little vulnerable, so Dave cobbled one together. Although that didn't take long, he then spent hours reaming out the electrical outlet in the cockpit, which was badly corroded. The effort was worth it, though, because the light made it much easier to spot *Windy Lady* on a dark night.

Towards the end of the first week, Dave made an early morning visit to the market, where he ran into Del and Joanne (*Limbo*). They were going to visit Cascades Waterfall, a local attraction not far from the city, and invited us to join them. While the day started warm and sunny, clouds were building on the horizon by the time we met them at the bus stop. We then focused too much on visiting and not enough on business, so failed to confirm the fare when we boarded the small van. It was too late to argue when the driver overcharged us at the other end. He then demanded even more to return and pick us up, so Del told him not to bother; we'd find our own way back.

Turning toward the trailhead, we saw two men standing at a nearby kiosk, and only then realized that we had to pay a fee to hike the trail. That was not something we were accustomed to doing, but it became a little more palatable when one of the men offered to join us. Speaking softly in English, he said his name was Robert. He was probably in his early twenties and taller than Dave, with a muscular build, very dark skin, and a mat of black, curly hair covering his head and jaw.

As we started up the path, Robert informed us that we would be following a stream up the mountainside and crossing it numerous times. So, when we emerged on the bank of a shallow stream a few minutes later, Del, Joanne and I removed our shoes, planning to continue in bare feet. As Dave wore sandals, he just carried on. The stream was about thirty feet across and the water only ankle deep, but the rock was smooth and slippery, so we stepped carefully.

We then followed a dirt path through the jungle, eventually coming out on the bank of a large pool, where a curtain of water spilled over limestone rocks on the lower side. As we waded across, we kept close to the edge where the current had washed away the mud. There proved to be four such pools along the trail, complete with freshwater fish; the water spilling from them dropped anywhere from eight to twelve feet.

A gentle rain began to fall as we climbed and it turned into a downpour that lasted half an hour. Cutting large, broad leaves, Robert offered them to us as makeshift umbrellas, but mine barely kept my glasses dry; fortunately, being wet in that climate wasn't a problem. Intermittent showers then continued for another two hours.

As we proceeded up the trail, Robert identified grapefruit trees, banana trees, pineapple plants, mangoes, and others that I didn't recognize. When we passed near a spring, he pointed out water taro plants. Farther on, he climbed a palm tree and cut down three drinking nuts; slashing the end off one, he handed it to Dave for the four of us to share. He cleaned up the other two and later gave them to us to take home.

The source of the stream was said to be a large spring gushing out of the rock on the mountaintop, and when we emerged from the jungle on the edge of the last pool, we could see several waterfalls leaping off the top of a cliff face high above. Falling in long, lacy veils, the water fell onto an outcropping of rock, then tumbled down in long cascades to a deep ravine, where it pooled and overflowed. It was a gorgeous spot, and I had trouble tearing my eyes away as I waded along the edge of the pool.

We continued following the streambed for another ten minutes, climbing up steps scoured into the rock, pushing against the rush of water spilling down. Savouring the moment, we stopped and took pictures of the spectacular scenery. Having thoroughly enjoyed the climb up, I was reluctant to turn back, but in no time at all, we were crossing the stream for the last time.

We then sat, putting on our shoes, and watched small groups of villagers emerge from the jungle and wade across in front of us. One group included three women and half-a-dozen younger children, while the older kids seemed to be on their own and came out in twos and threes. Burdened with wood, leaves, roots, and so on, they were returning to Mele Village and obviously had been foraging in the forest.

We continued on to the roadside stand at the trailhead, where we said goodbye to Robert. We then walked down the road towards the sea, where we came out on a black-sand beach with a beautiful view of an island resort. A sheltered spot, I thought, where we could possibly anchor. The rain ended as we approached a bus stop, and after a brief wait, we caught a ride back to town, paying only the usual fare.

By then, Del and Dave had discovered that they shared a curiosity about kava, the local drink, so the next day the four of us headed downtown to visit a kava bar. A cab driver directed us toward a modern-looking building, where we walked down broad, cement steps and entered a large, bare room. A row of kitchen chairs extended along the back and down the two end walls, while a small bar was located on one side of the entranceway. On the other side, a low, narrow sandbox stretched down the base of the wall.

A few men sat in the corner opposite the bar, so Joanne and I sat down on chairs in the middle of the back wall, while Del and Dave stood and looked around. We all watched when one of the men got up, walked over to the counter and handed over 100 vatu (about $1 US). After picking up his drink, he strolled over to the sandbox, stood for a moment facing the wall, then downed the contents and immediately spat into the sand. Returning his glass to the bar, he took a swig from a water bottle and spat again. A few minutes later, another man got up and repeated the routine.

I took careful note as the bartender swished the glasses back and forth in a tub of water and decided I'd pass on this experience; I'd already learned that my system didn't tolerate local bugs very well. I believe that Joanne had one drink, while Dave had four, and Del even more. Dave described the drink as tasting like chemically treated wood, then later added that his lips felt numb. Del kept testing his reflexes and maintained that he didn't feel a thing.

Del and Joanne returned to the marina before us as we had some shopping to do, so we didn't see what happened next. I understand, however, that Joanne had some difficulty getting Del into the dinghy and back onboard *Limbo*; reportedly, he stayed pretty close to the settee the following day.

During the ride back to town, I had confided to Joanne that small ants on the boat were driving me crazy. They were just everywhere, but particularly bothered me when I sat on the settee. They had a trail across the back, so would run over my hand, arm, or even the back of my neck. We figured they'd come aboard on the stock of bananas that Vaha gave us in Tonga, and then multiplied over the winter. A few days later, she tucked a small bottle of some product into our dinghy. I was forever grateful because it quickly brought them under control, eventually eliminating them altogether. (Although I did suspect the cockroaches that came aboard in Australia might have helped.)

A steady stream of cruising boats was then passing through Port Vila. Most stayed only a day or two before setting off to visit other islands. That was when we met George and Sarah off *Kemo Sabe* and, recognizing the boat name, found out more about the aborted fuel drop. Rather clumsily, I tried to convey just how much we admired them for their response to a difficult situation. We also ran into Don and Mimi off *Silver Cloud*, who answered our questions about the loss of *Camelot* on the reef off Viti Levu. Don then told us about *Rock Steady*, a Kiwi boat from the Bay of Islands; it was lost on the reef at Black Rocks, near Musket Cove.

When the weather started to improve, we began talking about visiting other islands, but didn't have the information we needed for route planning. All we really knew about the country was that malaria was endemic in coastal areas and it had many active volcanoes. So, we searched the stores for a cruising guide, and when we couldn't find one, Dave borrowed a guide and had it copied.

We learned that unlike Fiji or Tonga, the islands of Vanuatu were not behind a barrier reef. They were scattered across open water in a Y-shaped archipelago that was volcanic in origin. Because of a long history of isolation, unique customs had developed on some of the islands, making them popular cruising destinations. That was when we realized we were more interested in the present than the past. We decided to simply dayhop from island to island, learning what we could about the people as we traveled.

Having roughed out an itinerary, we obtained the necessary cruising permit and left Port Vila at 1100 on a Tuesday morning. As usual, the house batteries needed charging, so we motor sailed. After crossing Mele Bay, we followed the coastline around to the west and north; five hours later, we entered Havannah Harbor. The peaceful waters of this large, sheltered bay on the north side of the island had served as a staging point for US forces during WW II. Although I tried to picture what the bay had then looked and sounded like, my imagination failed.

We saw only one other boat as we traversed the length of the bay and anchored in a small cove at the far end. I then stood at the bow, coiling lines, and watched the lengthening shadows spread across the water while absorbing the quiet solitude. The three nights we spent there were gorgeous, the air cool, the early night sky filled with stars, and a half-moon then climbing up out of the sea. I was reminded of late summer evenings at a lake near home, and wouldn't have been surprised to hear the call of a loon. Here, however, the silence was broken only briefly by sounds carried on the wind; first, the distant putter of an engine, then the faint cry of a human voice.

While we were making breakfast next morning, we heard the soft murmur of voices drifting across the water. Rushing up on deck, we found a canoe sitting just off the port lifelines, with a man, woman, and two small children onboard. After greeting Dave, the man wanted to trade fresh vegetables for tinned meat or fish, only he didn't have any vegetables and could only promise to bring some the next day. Looking at the small family, Dave decided to risk a can of corned beef—and we never saw them again.

An hour or so later, some twelve to fifteen dugout canoes crossed the bay, carrying villagers from their homes on Moso Island to gardens that were located beyond the mangroves on the main island. About half of the boats stopped to check us out, one paddled by a young man whose loud singing arrived long before he did.

The villagers stopped again on their journey home that afternoon, when another youth hit Dave up for a liter of petrol. He refused as we carried only what we needed for the dinghy. A third youth sat in his canoe a few feet off the starboard rail and, seeming in no rush, chatted comfortably with Dave. He said that his name was John, he was twenty-one years old and unmarried; he also invited us to visit his garden.

The next morning, more canoes arrived but most stopped only briefly. John lingered on, however, and asked if he could see inside *Windy Lady*. When Dave gave permission, he called out to two of his mates, who quickly paddled over. Tying their dugouts to the toe rail, the men climbed over the lifelines. Taking them below, Dave showed them around, and after asking a few questions, they returned to the cockpit and climbed back into their own boats.

John repeated his invitation to visit their garden, and we made a quick decision to go with them, doubting that we'd find it on our own. As we hadn't put the motor on, Dave rowed the dinghy ashore, tying it up near a sturdy-looking outrigger canoe in the shade of the mangroves.

By then, John had started down a well-defined trail into the jungle, followed by a young man named Alex. Dave and I quickly fell into line, with two more youths bringing up the rear. Keeping to an easy pace, we followed the winding path underneath shady trees and through sun-drenched grasses.

The four men carried machetes, and I watched as John swung his blade up and slashed at certain trees along the trail, noticing that his mates did likewise. Watching the blades flash in the sunlight, I remembered that the ancestors of these young men had eaten "long pig"; they'd been cannibals. It wasn't difficult then to picture a man tied to a pole slung over the shoulders of natives following this very same trail. Telling myself that was a long time ago, I quickly brought my imagination back under control. Actually, it turned out not to be that long ago, as the last recorded cannibal feast in Vanuatu was held on the island of Malakula in 1969.

We walked for about twenty minutes, passing several intersecting trails, and eventually arrived at a large clearing in the jungle. Family members, including three women and six older children, were already at work, but they all came over to visit and have their pictures taken. John, meanwhile, proudly showed off their crops, from huge heads of cabbage to a pile of ripening tomatoes.

As we walked around the garden, John explained that individual family members were responsible for their own plots of ground, and had hand-watered the young plants until they were big enough to make it on their own. After burning off the brush initially, they had dug up the soil with shovels. Now crops were replanted in the same area, except for yams, which were relocated every second season.

For much of the year, the villagers drew fresh water from a narrow well, about eight feet deep, dug beside a small stream nearby. In the dry season, they caught a bit of rainwater but hauled drinking and cooking water from a stream farther around the harbour. Most of their vegetables were cash crops that they took to market in Port Vila on a weekly basis, including cabbages, tomatoes, lettuce, corn, green peppers and yams. They also grew bananas and papayas.

Feeling privileged to have spent the morning with the villagers, we somewhat reluctantly said goodbye and left them to their chores. Carrying a few tomatoes, lettuce and a yam, we found our own way back to the harbour. That night, John hit us up for some fishhooks and swivels as he paddled past on his way home.

Next morning, a clear, bright sky beckoned us onward, and we had the anchor up before 0800. I stayed at the bow as we motored across to the passage at Little Entrance, taking a long, last look at the quiet beauty of the harbour. Visibility was good as we passed through the channel, with a minimum depth of thirty-five feet.

CHAPTER 24

Vanuatu's Island Chain

As soon as *Windy Lady* was clear of the channel into Havannah Harbor, we stopped and raised the sails. Surrounded by quiet, sparkling waters under a bright, sunny sky, the morning seemed almost perfect. With NE winds of 15 kt, the bow was soon cutting smoothly through the ocean as we headed north-northwest. Before long, we passed near an active volcano on Nguna Island, where I searched the blue sky above the cone but saw no sign of steam. After we left the lee of Efaté Island, winds veered to the SE, bringing whitecaps and swell for over an hour, but Otto handled it all, leaving us free to relax and enjoy sun, wind, and sea.

Six hours later, four dolphins escorted us into Sesake Bay, an open roadstead on the northwest side of Emai Island. When still one-quarter mile off, we dropped the anchor in twenty-five feet of water. As we didn't go ashore,

I studied the beach through binoculars but saw no sign of a village. I then sat and admired the high, steep volcanic cone rising up into the sky on the west end of the island. My musings were soon disturbed, however, as *Windy Lady* began rolling slowly from side to side. Swell creeping in from the channel then grew higher and increasingly uncomfortable, disturbing our sleep overnight and causing us to gimbal the stove to cook breakfast.

We were underway by 0730, sailing off into another beautiful morning. With winds below 10 kt, all we had to do was raise the mainsail, pull up the anchor, and set the headsail. The winds increased to 10–15 kt farther out in the channel, bringing some clouds and putting a few whitecaps on the waves. By mid-morning, a line of squalls was passing behind us, sweeping across the sea from the west. I watched with fascination as the storms swallowed up the islands for thirty minutes at a time. When they moved on, the land emerged from the mist, like in a dream, and the sky again filled with sunshine.

The breeze was up and down during the eight hours we sailed that day. Dave had Otto working perfectly, but I enjoyed the sailing more when I stood at the helm and hand-steered. The sea flattened out as we drew into the lee of Epi Island, and we dropped the hook in the picture-postcard setting of Lamen Bay. Noticing that the three sailboats already at anchor were rolling noticeably, we tucked in as close to shore as possible.

After tidying the boat, we relaxed in the cockpit, enjoying the last hour of the day. Dave suddenly pointed to a number of outrigger canoes, just visible in the dusk.

They were crossing the channel to Lamen Island, and a few paddlers appeared to be using palm-fronds as sails. The swells then grew and the boat rolled throughout a second night; again, we gimbaled the stove to cook breakfast.

A few hours later, we lowered the dinghy over the side and Dave rowed ashore. Landing in a three-foot surf was another first for us, and the waves tossed the dinghy about roughly before throwing it up onto the black sand beach. Thankfully, it stayed upright with no damage done except to our dignity. Pulling it well up on shore, we then strolled through the sands for half an hour, watching the surf roil the water for some twenty feet out, while enjoying the feel of warm sunshine and soft breeze.

The long beach curved around the bay, and we soon noticed that the sands ahead were no longer black. In fact, both ends of the beach were blocked by banks of coral rubble that had washed up out of the sea; in places, the steep banks were ten feet high. Leaving the beach, we crossed a flat, grassy meadow, which Dave recognized as an airstrip only after picking up a discarded boarding pass. We then circled around through the adjacent village but saw no sign of activity.

We continued on to a restaurant that sat across from the spot where we'd left the dinghy. Although it appeared to be closed, we stopped to admire a tree fern carving and tam-tam (slit drum) displayed in the immaculate grounds. Dave saw a man emerge from the building, so walked over and asked if we could get a meal. The fellow shook his head, saying, "No, sorry, not without notice."

We walked on toward a large building visible in the distance, but partway there were intercepted by half a dozen young women cutting across the field. The girls appeared to be in their mid-teens and spoke English, so we were able to satisfy each other's curiosity. We told them about our travels and where we'd come from. They explained that they were students, and that the building ahead was a high school, or more accurately, a boarding school.

Some 150 young people attended the school, many of them coming from the Shepherd Islands to the southeast. After completing a four-year curriculum, each student would earn a grade ten certificate. Impressed by their fluent English, we asked the girls about their other languages. Each one had first learned the language of her village, then Bislama, followed by English and now a little French. At that point, Dave interjected, "Parlez-vous francais?" Something about his accent seemed to tickle their funny bones because they all laughed delightedly.

Having thoroughly enjoyed our visit, we said goodbye and returned to the dinghy. Although the swells were higher than before, we had no problem launching it back into the surf. When we approached *Windy Lady,* however, she was rolling far over onto her sides. As it seemed she would hit us if we came too close, we stood off and Dave studied the situation.

Deciding it was all a question of timing, he positioned the dinghy so that when the hull rolled toward us, I could grab the bottom rung of the rope ladder that hung from the toe rail. I held on as the hull rolled back, which pulled us in alongside. When it started rolling towards us again, I launched myself up the ladder while pushing the dinghy away. I took the end of the painter with me so that, once on board, I could pull the dinghy in and Dave could make the crossing.

The sea had calmed considerably later that afternoon when we visited with Del and Joanne onboard *Limbo*. There was little drama as we climbed down into the dinghy, and even less when we boarded the other boat, as Del had put out stabilizers to reduce the rolling motion. The swells grew again after supper, eased for a few hours before midnight, and then settled back in for the rest of the night.

Having put in a third uncomfortable night, we decided to move on. I regretted not having seen the resident dugong, but the early morning appearance of a large sea turtle more than made up for it. The turtle swam in amongst the six boats then at anchor, rolling over on its back and splashing about with large flippers.

That day was quite stormy, and with winds behind us, we set only the headsail as we started up the wide channel stretching between the islands. Apparent winds of 10 kt increased to 15 kt about midday, and we each took a turn at the helm during the six-hour crossing. A halo of cloud hid any steam that might have been venting when we passed close to the active volcano on Lopevi Island. Ultimately, we so enjoyed the day that we sailed past our next anchorage and turned back into gusty 25-kt headwinds.

The anchorage at Craig Cove on Ambrym Island

proved to be nothing more than a slight indentation in the shoreline. It looked rather dreary, with dark jungle creeping out to meet the black lava rocks scattered across the beach. As we weren't far from the channel, we tucked in as close to shore as possible before dropping the hook in forty feet of water. Although I suspected we were in for another bumpy night, I woke only twice when the boat rolled sharply.

Not long after we arrived, a leaky, outrigger canoe eased up beside *Windy Lady*. It contained five young lads, maybe fourteen to sixteen years old, one of whom bailed constantly. Our attempts to speak to them were disappointing as the boys spoke only French. Both Dave and I had studied the language in high school but never used it; now, I barely remembered anything and was reduced to hand motions. Dave did somewhat better, managing to buy two papayas from one boy who knew a few words of English. He also figured out that they had no vegetables to sell because the lava was too new.

All efforts at communication then stopped, and the boys continued to sit alongside for the longest time, not even speaking amongst themselves. When they left, a smaller dugout canoe appeared. This one carried three young girls, probably about the same ages, one of them held a year-old baby on her lap. They too spoke only French and we couldn't communicate at all, just sat looking helplessly at one another, until they too returned to shore.

The next morning brought clear, sunny skies with white horses leaping out in the channel; it looked like another day of strong winds. Because we were so close to shore, I spent a couple of hours checking the GPS, making sure that we weren't moving. We went ashore just after midday, stumbling through black sands and over lava rocks as we carried the dinghy up from the beach.

We bumped into an Australian couple, Bill and Judy on *Hacienda*, who had arrived that morning and were then ashore. Joining forces, we walked down a dirt track toward buildings partially hidden in the trees. A cloud of flies buzzed annoyingly around our bare legs, attracted I assumed, by the garbage that was strewn about. Earlier, when I was looking through the binoculars, I'd seen a woman dump a basket of waste over the rocks at the water's edge.

A few minutes later, we found ourselves in front of a long, roughly finished building displaying a "Co-Op" sign. Curious, we went inside and found a wide variety of goods on the shelves; however, about 20% of the space was allocated to soft drinks and liquors. We then strolled through the nearby village, passing by small homes with thatched roofs, woven mat walls, and neatly maintained grounds. A few chickens wandered about, as did a whole swarm of pre-school kids, while several men loaded sacks of what we took to be copra into a battered pickup truck.

I saw very few women and assumed they were tending bush gardens or foraging in the jungle. We figured the older children were probably at school, while students at high school level would have left the island to continue their education. I then spotted three pickup trucks driving away from the village, each with a full load of men sitting in the back. I had noticed a man in a small powerboat leaving the bay earlier, but was starting to wonder what men did to keep themselves occupied.

From radio reports, we had learned that the government's attempt to develop the economies of the outer islands had ended in the early 1990's, when copra prices crashed. We also heard that some families could no longer afford to send their children to school. For a long time, I had puzzled over policies that made subsistent farmers dependent on cash crops. If unsuccessful, as in this case, what could they do? And without an education, what hope did these young people have for the future?

The teenagers we'd met the day before, sitting silently in their canoes, made such a sharp contrast to the girls on Epi Island, whose faces had glowed with confidence. But even those young women would have to leave their villages for larger centers, like Santo and Port Vila. What then happened to the traditions and support systems of these small villages? It seemed to me a conundrum of our times that success for current generations meant destroying values developed over centuries.

After spending two nights in Craig Cove, we set off once again. I felt depressed as I watched the shoreline recede, thinking that these young people seemed trapped. I told myself that maybe I was wrong; maybe I was over-reacting because we hadn't been able to talk to them. Then I remembered the girl in the canoe with the baby, and a wave of sadness swept over me. I'd seen her and the child loitering in the store, then later in the village, and assumed it was hers. I also recalled the shelves of pop and liquor in the store and wondered who bought it.

Because of southerly headwinds, we motor-sailed most of the way to the island of Malakula. The three-hour crossing was again extraordinary, with sunny skies and sparkling blue waters. We paid extra attention as we approached the channel into Port Sandwich, concerned about a shallow reef that extended out from one side. In fact, we were too concerned, so nearly went aground on the opposite side, with less than four feet of water beneath the keel when the bottom leveled off.

The channel opened up into a large, protected bay with a high, wooden dock located near the entrance. As we came around, we could see, "SHARK" written in large, red letters across the front, which certainly stifled any desire for a swim. We dropped anchor at 1320, and for the first time in nearly a week, *Windy Lady* sat quietly in calm waters.

We tidied the cockpit, then went ashore and followed a narrow dirt track into the jungle. According to our guidebook, it led to the coastal town of Lamap, some distance away. We walked only a few miles that day, happy to get some exercise, but also eager to learn what we could about rural island life.

Plantations of coconut palms lined the road, with the tall, slender stems and leafy tops swaying gracefully in the breeze. The debris underneath appeared to have been piled and burnt, with small circles of ash all that remained. A few cattle grazed close to the road, and every so often, we passed a sheet-metal-roofed copra dryer sitting idle, back under the trees. We also walked by a few individual homes and a small village; the surrounding grounds appeared neat and clean, and were decorated with flowering and variegated-leafed shrubs.

As we approached the dinghy on our return, we met a slim, middle-aged woman named Mary, who lived nearby. We were relieved to hear her speak English, as we'd read that people in that area spoke only French. A child, about eighteen months old, played fitfully beside her, his small, naked body covered with sores that looked a lot like chicken pox. She told us he'd had a "slow" fever.

When asked where she lived, she pointed towards a small metal-clad shed situated about three feet from a copra dryer near the wharf. She shared it with her husband, child, and an older woman. With metal on roof and sides, one door, and no windows, it looked like it provided minimum shelter. I later noticed that she spent all of her time working outdoors. I assumed she probably had a bush garden too, because when Dave asked about fresh vegetables, she said she'd bring tomatoes and cucumbers the next day.

That evening, I sat outside as usual, watching the light fade from the sky. As I absorbed the quiet serenity of the bay, I noticed an outrigger canoe emerging from the shadows. When it drew nearer, the paddler called out a greeting and asked if I spoke French. When I confessed that I didn't, he continued in English. He then sat alongside for maybe ten minutes, and we spoke about the voyage from Canada.

Looking at him closely, I figured he was maybe thirty years old. He wore only red knee-length shorts and sat on a board placed across the gunnels of a very narrow dugout canoe, his legs crammed together inside. As we talked, he sculled constantly with his paddle, causing muscles to ripple across his arms and chest, while his deep laugh pealed out freely, seeming to float across the water.

He was a very impressive individual, in both manner and appearance, and I was quite taken by his laughter. But when I tentatively asked permission to take a picture, he shook his head. When he later disappeared into the deep shadows that had settled over the bay, I sat and wondered about his education, his obvious fitness, and especially his joie de vivre. How did he fit into a subsistence existence in Port Sandwich?

Later, well after dark, I decided it was safe to take a bath in the cockpit, so undressed and started to wash my hair. As I wiped the soap from my eyes, I heard a male voice call out a greeting and ask, "You want lemon?" Sinking to the floor, I crouched down behind the cowling and returned the greeting as graciously as I could, adding "No, thank you."

Next morning we did a few chores, using bucket and plunger to wash and rinse a load of clothes, which I hung on the lifelines to dry. As the rudderstock was leaking again, we pulled the port berth apart and Dave managed to tighten the lock nut a little more. By 1300, the berth was remade, clothes were dry, folded and stowed, and we were free to go off and explore.

As we left the dinghy, we saw Mary and two other women sitting on the grass under the trees. In their midst was a pile of bananas that had been split in half lengthwise. The women chatted as they scraped out the pulp, which they threw into a big pot. A large pile of folded leaves sat on the grass nearby, and Mary explained that they would wrap the pulp in the leaves and bake it underground. A fourth woman arrived at that point and said they were making lap-lap, or island pudding. I figured that was probably something else I would have to avoid.

Once again, we set off on the road to Lamap, enjoying the sights and sounds along the way. As we passed through one of the villages, we stopped to take pictures of several children with their mothers. The road wound through more coconut plantations, passed by a men's meetinghouse guarded by two tree-fern carvings, then came to a kava bar, with pole benches inside and sheet-metal coverings outside. Eventually the road forked, and as both routes looked the same, we followed the one to the right.

A half-mile farther on, we heard a voice call out and turned to see a native man working at a copra dryer about 100 feet off the road. As we walked beneath the palm trees towards him, I saw the blue haze of smoke rising up against green leaves, and then orange-red tongues of flame darting out below. The man hollered again, saying cheerfully, "If you're going to Lamap, you've taken the wrong road."

Dave laughingly admitted that we were, but when he showed an interest in the dryer operation, the chap seemed happy to talk. Continuing to tend his fire, he explained that after the coconuts fell from the trees, they collected and husked them. The nuts were then split in half, exposing the white meat inside, and placed on the drying rack above the fire. (I later read that thirty coconuts provided about ten pounds of copra, which is the source of coconut oil.)

This fellow, who had a husky build and looked to be in his thirties, now asked, "Are you thirsty?" He then quickly climbed thirty feet up the slender trunk of a nearby coconut palm and knocked down several green drinking nuts. After shinnying back down, he skilfully trimmed the end of one with a machete, then slashed straight across it with one cut, leaving an opening about an inch and a half in diameter from which to drink. He presented it to Dave and then prepared one for me. All the while, I stood in wide-eyed wonder, admiring both his physical dexterity and his skill with the machete.

When we set off once more, we backtracked to the fork in the road and took the route not yet traveled. Soon, we were at the old town site of Lamap, which at one time had been the center of French Administration on the island; it had been vacant since independence in 1980. We wandered up and down streets that had deteriorated into little more than country lanes lined with shabby-looking buildings.

After a while, we came to a large grassy area beneath a huge banyan tree and saw a small building set back in the shade. At the entrance to its long walkway, two twisted pieces of driftwood, each painted with a face, identified it as a kava bar. From a cliff top, we then stood and looked down on a broad reef that ran along the shore, and admired magnificent views out across the channel towards Ambrym Island.

While hiking back to Port Sandwich, we started

meeting women returning from their bush gardens. Sometimes a woman was alone, sometimes there were two or three together, and we came across several resting in the shade. They were all loaded down with coconuts, corn, leaves, roots, or pieces of wood that they carried in dirty pieces of material thrown over their shoulders. One woman also carried a shovel. I then recalled seeing a few women on the road the day before, although it had been earlier when we returned from our walk.

I pondered over my impression that men and women lived quite separate lives. The women seemed to be away from the village most of the day, tending gardens or foraging in the bush, while the men had been most noticeable sitting in the backs of pickup trucks. They obviously looked after the copra crop and probably fished, although we hadn't seen much evidence of that.

Mary had fresh cucumbers, tomatoes, and a papaya waiting for us when we reached the harbour. Dave rowed out to the boat and brought back two cans of corned beef in trade. As things went, that was probably too much, but we had no use for the canned meat and felt the fresh vegetables were worth it.

Not long afterwards, another canoe approached *Windy Lady*. The young man paddling it was soon explaining that he was going to pick up his wife, who had spent the day working in her garden. He also spoke English and when I asked him whether he spoke French, he just shook his head. I noticed that his dugout canoe looked more finished than others I'd seen, with wooden strips added to the sides, raising the gunnels. What really caught my eye was his paddle, however, as it was nicely tapered and a perfect teardrop in shape.

Friday morning brought strong southerly winds, and we were content to spend the day onboard. We discussed moving on to the next anchorage, eight miles away, as it was supposed to be safe for swimming. However, we'd heard several reports of shark attacks that season, and I didn't think the move would make me feel more comfortable in the water. About 1400, a small inter-island freighter poked its bow into the harbour and tied up at the dock. For the next four hours, we watched a flurry of activity ashore, and some of my questions about what men did to keep occupied were answered.

First, a crew of about a dozen men offloaded various bales and barrels, toting them from the deck of the freighter down the gangplank, depositing them in a pile on the dock. The goods were then loaded into the back of a pickup truck. When piled high, it disappeared into the jungle, and a trailer pulled by a farm tractor took its place. The two vehicles alternated about every thirty minutes until the supplies were all gone. The ship's crew then loaded the sacks of copra, cacao, and empty barrels that waited nearby. (Cacao pods grow on small tropical trees, and according to Mary, were used in making both chocolate and coca cola.)

We were still sitting outside when an outrigger canoe appeared out of the lengthening shadows. This time the paddler was a fairly small man with grey in his hair and not too many teeth. He was soon standing in his canoe, gripping the toe rail with one hand and talking to Dave. I wondered whether he stood in an effort to bring the level of his head closer to ours, as I'd read that was a concern in some cultures. But I soon suspected that he considered me rude for asking questions and therefore ignored me.

He seemed inclined to stay and talk to Dave, however, telling him that he had five children, with the eldest living in Port Vila. When Dave remarked on the fact that he spoke English, he nodded and stated gravely, "Yes, I'm Presbyterian." He went on to explain that the people who lived on this side of the harbour were "Catholics Roman", so did not speak English. Where he lived, on the other side, they were Presbyterian and did. As the shadows deepened, he waved his hand toward a 5-gallon plastic jug in the bow of the canoe, and said that he was on his way to the freighter to get kerosene. I kept watch as he paddled over to the wharf, and a few minutes later, saw him climb back into his canoe and disappear into the night.

The freighter departed at 0530 next morning, passing close by *Windy Lady* on its way out. Awakened by the throbbing of the diesel engine, followed by flashing lights and a bow wave that rocked the boat, we got up, made coffee and mulled over our plans. I paged through the cruising guide, reviewing alternate destinations, but in the end, we decided it was time to return to Port Vila. The forecast was calling for sunshine and southerly winds of 20 kt, but we suspected they could easily strengthen. As we would be pushing into headwinds, we decided to cover the distance as quickly as possible, and made plans to sail through the night.

CHAPTER 25

Port Vila Again

We pulled up the anchor at 0730, and Dave then turned the bow towards the harbor entrance. I stayed on the foredeck, waving goodbye to Mary and another woman, who watched from shore. Once we were clear of the entrance, we put two reefs in the mainsail, pulled out the headsail, and headed east. Initially, winds were on the beam at 10–12 kt but gradually they increased to 25; with sunny skies and sparkling waters, the sailing was superb for the first two hours.

Once we had a clear route south, we tacked, which had *Windy Lady* beating into SE winds of 20 kt and plunging into lumpy eight-foot seas. The swells grew higher as the day progressed, and by late afternoon, she was falling off waves. Each time she did so, the bow smacked loudly against the water, sending a shudder running through the boat. Fortunately, we drew into the lee of a small island before conditions became too bad.

At 2000, we were clear of the island, winds were

gusting at 20–25 kt, and seas were up to ten feet. *Windy Lady* was falling off waves again, but now she hit the water so hard that I could feel the shudder in my own core. Hoping to reduce the beating she was taking, we put a reef in the jib, trying to slow her down. But seas remained extremely rough overnight; the GPS quit working at 0200 and neither of us were able to sleep.

By 0430, winds were strengthening and seas were up to twelve feet. We were then approaching the western side of Efaté Island, so Dave started the engine; setting the throttle at 2300 rpm, the boat inched along at two and one-half kt. By first light, we were nearing the southwest corner of the island, and winds were hitting 30 kt. We tracked too close to shore as we rounded the point leading into Mele Bay, and huge waves hit *Windy Lady* on the beam, pounding her brutally and rolling her over so far, that the masthead must surely have hit the water.

With the boat reaching extreme positions on either side, all we could do was hang on; meanwhile, everything that wasn't nailed down inside the cabin went flying. I don't know how long it took to clear the maelstrom off the point, but two hours would pass before seas were noticeably calmer. We finally dropped the hook in the harbour at Port Vila at 1100 hours, having logged 262 nm in fifty-six hours of local sailing.

I was so tired that I ached all over, but Dave was fussing about, so sleep wasn't an option. With a layer of salt residue and dust coating everything inside the cabin, I spent the afternoon cleaning. Next day we checked in with Customs, turned in our cruising permit, and started to think about the passage to New Caledonia.

As the headsail had a small tear, we found a sailmaker and dropped the sail off for repairs. Dave also looked for a service center for the GPS. When he couldn't find one, he ordered a hand-held unit from West Marine, knowing that delivery would take at least a week. We had originally discussed doing so while in New Zealand but procrastinated, so now paid $50 for a phone call, $60 for airfreight, and $150 for the unit (in US funds, of course).

Eventually, he located a marine electronics store and took the mushroom antenna in for testing. While the unit checked out okay, it took an hour to pick up satellite signals when he reinstalled it. After he'd made numerous trips back to the shop, the tech finally made a trip out to the boat and discovered that the pin inside the cable fitting on the antenna was only making intermittent contact. He figured that someone or something had yanked on the cable. Although he didn't have a replacement part, he did fix it.

About then, we heard that a group of dancers from northern Ambrym Island would perform a *Rom Dance* in Port Vila. The dance was originally part of the all-male grade-taking ceremony held on the island every August. Traditionally, there were more than thirty grades in the hierarchy governing the lives of men on the island. Apparently, a man rose through the ranks by holding parties and giving away all his wealth. (It was rumored that women did the work that created the wealth.)

This was a rare performance by the group, and I was

excited at the prospect of seeing it. The venue was within easy walking distance of the marina, and we arrived early, paying a fee of 1000 vatu each (about $10 US). That amount gave us access to a wide, grassy field at the Chiefs' Nakuma and allowed us to take still pictures. (For a video camera, the fee was 3000 vatu.) We strolled around the field for a while, then stopped and inspected the woodcarvings at the handicraft stalls.

We weren't surprised to see two pigs tied up beneath the trees nearby, as the animals were highly valued in traditional culture. A boar's tooth was a status symbol; one that made a complete circle identified the wearer as a chief. We now learned that the smaller pig was the price paid for using the facilities for the afternoon.

As more tourists crowded into the area around the handicraft stalls, I felt uncomfortable and moved off to the side. But when I turned away, I came face-to-face with a naked black man standing about a foot behind me. The man was slim, not much taller than I was, and wore only a penis cover. Completely discombobulated, I desperately looked around for Dave but he was still examining the woodcarvings. I managed to edge out of the crowd and, while waiting, noticed several similarly dressed performers moving nonchalantly amongst the tourists.

The spectators soon settled themselves around the edges of the field, and the performance began with the beat of a tam tam (slit drum). The chanting of male voices followed, as a tight circle of dancers shuffled forward. The sight was extraordinary as, at first glance, the performers looked like moving haystacks. Costumes made from the long, dried leaves of banana plants draped their bodies, covering all signs of arms and legs. Then I saw the elaborate painted facemasks and tall bearded headdresses that concealed their heads and hair.

Stamping their feet, the dancers moved slowly towards us, and now we could hear the rattling that came from the long funnel-shaped wands that each one carried. As they moved farther into the field, they began to spread out, revealing glimpses of men hidden in their midst. This second group of men then danced forward, revealing themselves as the singers, and one of them kept time by knocking one stick against another.

The singers were naked except for penis wrappers, which attached to belts that encircled their waists; a few had added leafy branches to the backs of their belts. The lead dancer was an older man with grey hair and beard; he carried a spear and wore a boar's tooth bracelet on each wrist. A third tooth, one that formed a complete circle, hung from his neck. (When I took his picture earlier, he had closed his eyes while posing.)

After a time, the singers withdrew and the masked dancers again became the center of attention. A tam tam had been set up on the edge of the field, and the performers now whirled fiendishly across the grass, shaking their rattles vigorously. As the costumes rustled ever louder, pieces of leaves fell onto the grass, where an errant breeze picked them up and sent them dancing through the air with a life of their own.

My scalp was starting to tingle with the eerie, otherworldly quality of the scene, but then a dancer twirled and the flash of red shorts broke the spell. Soon after, I noticed a pair of Nike running shoes on the feet of another performer. It was then apparent that some of the dancers didn't know the routine very well, as a couple of older men watching from the sidelines continually pushed them into position.

After about an hour, the focus shifted from the dancers to the larger of the two pigs we'd seen earlier. A man, pulling on a long tether attached to a front foot, dragged it onto the edge of the field. Another man then approached and, in one swift, powerful motion, swung a club back, up, and down. I didn't see or hear the club strike but closed my eyes when the pig's squeals echoed across the field. The pig continued squealing as three different men struck it numerous times on the skull, and finally it fell silent.

Each dancer now came forward and touched the pig's body with the end of his wand. They then spread out across the field, sweeping away any spirits that might have lingered after the dance. Traditionally, even the costumes were destroyed after a performance, as it was feared they might retain the energy of spirits represented by the dancers.

By then, all the interest and enjoyment I'd felt in watching the performance had fled. The brutal killing of the pig had shaken me, leaving me numb. Looking around the field as people departed, it seemed that a quietness had settled over everyone.

I accompanied Dave back to the handicraft stalls where he quickly came to terms with a vendor. Faced with the prospect of shipping goods back to Ambrym, the man had already dropped his prices. Dave left carrying a heavy, three-foot-long hardwood slit drum with a beautifully carved head. Only then did he tell me about a conversation he'd had with a native man, who had stood beside him for part of the performance. Turning towards him, the man had casually observed, "It's not that long ago, you'd have been lunch!"

As we continued to wait for the GPS to arrive, we began thinking about route planning beyond Australia. With the expiration of the British lease in Hong Kong, we wondered whether there might be policy changes that would affect cruising boats. As the Chinese Consulate was within easy walking distance, we stopped and inquired. The Consul obviously wasn't used to that type of question, but assured us that it would be business as usual, with the only obvious change being the new Governor.

I was now spending my time studying charts, pilots, and tide tables, preparing for the crossing to New Caledonia. Our thirty-day visas then expired, so we renewed them for another month. When the GPS finally arrived, we cleared with Customs and Immigration on the same day.

The following morning, we went ashore to make a few purchases and ran into a couple who'd just attempted the crossing and turned back. They assured us they'd never turned back before, but strong winds on the nose and rough seas had been more than they wanted to handle. Despite continued reports of headwinds from boats at sea, we left for Noumea the next day.

CHAPTER 26

Passage to New Caledonia

The crossing to New Caledonia would take only three days but presented new challenges. First, in order to make it through a pass in the Loyalty Islands, we had to maintain an average heading of 195 degrees (SSW) for over forty-eight hours. At the same time, we would be beating into strong SE winds and SSE seas that would push us westward. We then needed to be in position off Havannah Pass at first light the following morning, so we could enter through the reef on a flood tide. If all went well, that would have us in Noumea later that day.

We raised the anchor at 0945 on October 5, putting two reefs in the mainsail before leaving the shelter of Port Vila's harbor. Minutes later, we were shutting down the engine and pulling out the headsail. With SE winds at 20 kt, *Windy Lady* then charged into oncoming swells; soon she was making five kt and pounding into rough six-foot seas. Right from the start, we noticed a westward drift.

When winds veered to SSE that afternoon, the sea noticeably eased but the respite didn't last long. Gusts of 25 kt soon had *Windy Lady* racing through the water at six kt, and before long, she was twisting and turning violently as she again plunged into lumpy six-foot seas. Dave put three reefs in the headsail during his 2000 watch, slowing her down to four and one-half kt, but still our berths pitched and rolled, so we couldn't sleep.

Winds increased to 30 kt around midnight, and we then measured our westward drift at one nm/hour. Dave checked our position at the start of his 0400 watch and realized that we had no hope of making the pass. So we tacked, turning to a heading of 070 degrees and sailed eastward for four hours. When we turned back on course, our longitude reading on the GPS was the same as it had been twelve hours earlier (167°54'E). While we gained fifteen nm of easting, we were five nm farther from our destination, so lost ground in the process.

Upon turning the bow southward, SSE winds of 20–25 kt drove *Windy Lady* forward at five and one-half kt, and she careened into swells that were eight to ten feet high. I stood at the helm throughout my morning watch and hand-steered, watching waves washing over the bow and spray flying high over deck and cockpit. Conditions were hard on both boat and crew, and at noon, after twenty-six hours at sea (Day 1), we logged 117 nm.

The seas eased for a few hours early that afternoon, but then 30-kt gusts from the SE again pushed them up to ten feet. The strong winds also pushed us westward, so at 1600, we tacked and turned to a heading of 090 degrees. At 1845, I noticed the lights of a ship bobbing about on the horizon, and Dave tracked it on radar for thirty minutes as it passed five and one-half nm astern. When we turned back on course at 2000, the numbers were almost identical with our earlier effort; we gained fifteen nm of easting, lost five nm to our destination and were one minute of longitude off our previous position.

Sea conditions that evening were the roughest we ever experienced. When Dave was standing at the counter preparing supper, he was flung backwards across the galley and ended up sprawled on a bench at the table with paring knife in hand. Something similar happened to me a few hours later, when I was standing at the chart table, filling in the journey log. I must have eased my grip on the grab rail when I leaned over to read the barometer, and the boat then lurched violently. I flew backwards, hitting the back of my knees on the coffee table and curling up over top of it. I ended up sitting in Dave's lap on the settee.

About midnight, the winds backed a few degrees and dropped to 20 kt. That allowed us to steer our course while maintaining a boat speed of five kt and proved to be the turning point in the voyage. We took one reef out of the headsail, and as seas grew quieter, managed to get some sleep; by 0800, seas were down to three feet. Winds continued to gust up to 25 kt during my morning watch, but I had no problem staying on course. In fact, with a bright, sunny sky and sparkling waters, the sailing was most enjoyable. At noon, we recorded 116 nm for Day 2.

After sailing 230 nm, we arrived at our waypoint off the Loyalty Islands pretty much on schedule. We were through the pass before dark, and with quieter seas, the boat speed averaged six kt for the twelve-hours ending at midnight. By then, winds were down to 15–20 kt, but still we reefed in the sails, slowing the boat down, as we didn't want to arrive at Havannah Pass much before daybreak. I was actually relieved to do so, as it was a bit unnerving to race through the darkness while passing close to flashing lights that brightened the waters around nearby shoals. When we reached our waypoint at 0612, we had three reefs in the main and five in the headsail.

Just after low tide, with the sun not yet over the horizon, we started in through Havannah Pass. The channel was notorious for cross currents and hidden shoals, but we had the engine running and were coming in at the best time, the start of the flood tide. Dave was on the helm and steered us through safely, although my stomach knotted up a time or two when currents caused the boat to slew sideways. I wasn't too happy either, when a rainsquall reduced visibility to zero for forty-five minutes.

He had earlier entered nine waypoints into the GPS to guide us through the forty-mile route into Noumea, so turned on the autopilot once we were through the entrance. With only a few adjustments, it then guided us past the islands and reefs adjacent to the channel. After motoring for eight hours, we tied up at the marina at 1400.

Overall, we felt good about the passage, although our rhumb line route of 320 nm had stretched to 382 nm logged. When I wasn't fretting about sea conditions, I had delighted in the gorgeous, sunny skies and brilliant, star-filled nights. A pair of dolphins had accompanied us briefly when we came through the reef, reminding me of how little sea life we'd seen that season. There had been a pod of maybe six small whales off Malakula Island, and one large, playful sea turtle at Epi Island.

I stayed on deck, putting the boat to bed while waiting for the authorities to arrive. When I saw a Customs Officer walking up the dock, I first noticed the large handgun on his hip, and then greeted him with a cheery, "Bon jour!" Beaming, he responded in kind and added, "Parlez-vous francais?" When I shook my head and said that I didn't, he grew visibly excited and exclaimed, "But you are Canadian, surely you much speak French!" As I continued to shake my head, he muttered, "You should be ashamed! You are Canadian and don't speak French!"

After that exchange, the Customs Officer refused to speak to us. He climbed onboard *Windy Lady*, then stood beside Dave and rudely thrust papers under his nose, pointing to whatever needed signing. The few questions he asked were in such poor English that I couldn't understand a word. Fortunately, Dave knew the routine by then and was able to respond appropriately. When the officer left (he was white), the Immigration and Agriculture Officers came onboard (both natives). Neither of them spoke much English, and while surprised that we didn't speak French, at least they weren't offended.

That was our introduction to the international world of French-speaking countries, where officials apparently believed that Canadians spoke French. American cruisers we met didn't have the same problem, but we went through similar treatment when we checked out. I can only assume this occurred because our federal government promotes the country as francophone, although less than 25% of citizens acknowledge French as their mother tongue (1996).

Actually, I was required to study the language for four years in a public school system that provided teachers who did not speak it. Although I studied conscientiously, the fact was that without any practical use I lost even my reading skills. I now believe that my time, as well as taxpayer money, was wasted on that part of my education.

The next morning, we eagerly started exploring Noumea and found our way to the wet market, arriving just before it closed at 1100. We hadn't yet found an ATM, so couldn't buy anything, but enjoyed wandering through the complex, which was something of a culture shock after our time in Port Vila. Two sections contained fresh fruits and vegetables, as well as a bakery and fish market. In a third area, shells and trinkets were attractively displayed on tables, while a band played music nearby. There were no piles of roots or coconuts on the floor, and no sign of the oranges that had started to flood the other market. Even more, there were no natives sitting on the floor and all the sales people were white.

We continued into town, exploring up and down hilly streets, occasionally pausing to visit a store. Noumea appeared quite European; we noticed only a few dark-skinned people and saw no sign of native dress. Posted shop hours were 0700 to 1800, with a three-hour closure at midday, from 1100-1400. We ended our excursion with a late lunch at a French restaurant, discussing our first impressions while enjoying the novelty of a glass of wine. The truth was that we were tied up at a marina for the first time in seventeen months, and just starting to appreciate being able to step off the boat onto a dock.

With the weather sunny and warm, we made a point of visiting some of the city's many attractions. First, we boarded *The Little Train* for a tour of the downtown streets. We then stopped off at the Botanical Garden and wandered through the grounds for hours, ending up at an unusual cactus garden. Along the way, we visited numerous aviaries and were introduced to a flightless bird called a Kagu, which is native to the country and a national symbol.

I was frustrated at the Territorial Museum, however, as the storyboards were all in French. The only thing I learned was that native people were called Kanaks, which struck me as surprisingly similar to Canucks, which Canadians are sometimes called. Dave fared a little better as he recognized a word here or there, so found a few of the exhibits of interest.

The aquarium at Baie d'Citron more than made up for my previous disappointment. While the facility was small, the exhibits were well planned, with natural lighting and fresh seawater circulating in the tanks. We saw many colorful fish, as well as lobsters, seahorses and sea snakes. One small fish, called "the bulldozer", was digging a nest in the bottom of the tank by sucking up mouthfuls of gravel and spitting them out to the side. Another species featured was the Nautilus, a large, deep-water mollusk that grows up to eleven inches in diameter.

Of even more interest were the phosphorescent corals on display in a darkened section of the facility. These exhibits were breathtaking and our time there went much too quickly. One of the corals had a dim, slender stalk with a bright button of light at the top; a second appeared bathed in a silvery glow; a third resembled a display of crystal. Other corals were bathed in either green or orange hues coming from what looked like threads of colored light woven into their surface.

On the long walk back to the marina, we stopped and joined spectators enjoying fun day activities at a small beach. Four blindfolded individuals were then being helped into two small dinghies. Officials placed buckets over their heads and provided them with plastic dustpans. Guided by voices calling from a diving raft about 100 feet off the beach, the contestants set off, paddling madly with the dustpans. Not surprisingly, the two teams were soon going in circles, and neither completed the course.

I noticed then that most of the onlookers were men, and they weren't watching the boats. Unobtrusively, all had their eyes peeled on nearby sunbathers, where three young, Eurasian beauties were displaying most of their assets. Wearing only colorful thongs that accented the curves of buttocks, their slim, brown bodies were otherwise bare. As I watched, one woman got up and sauntered around, giving everyone an eyeful.

Although we normally preferred to explore on our

own, our time was growing short, so we decided to take a guided tour of the island. Dave then signed up for the most expensive one he could find, which was a full-day affair. Driving a comfortable eight-passenger van, our guide picked us up at the marina at 0800. He proved to be a forty-something Englishman, who loved the French mystique of sexy women, hot-blooded men and sophisticated culture.

Our companions were two Australian couples from Brisbane and Sidney, and two single women, one from Sydney and the other from Paris, France. The girls were both very attractive and made a striking contrast. Penny was slim and blond in the outdoor Aussie manner; Gail's clothes were fashionable, her dark hair smartly styled, and face carefully made up.

Our guide provided an exuberant patter as he drove north out of the city and across the central mountains towards the east coast. He then turned onto a narrow one-lane road, where he explained that the direction of travel was controlled and changed every hour. As we followed the road north through rugged mountains, we passed numerous nickel mines sitting high on ridgetops, where the ore bodies were located. Road networks ran everywhere, and the open-pit mining process had removed whole mountaintops. Land erosion was an obvious problem, and our guide informed us that mine waste had polluted many rivers.

After leaving the one-lane road, we drove on until midday. We then stopped at a small restaurant, where we enjoyed a delicious meal featuring delicately seasoned pork chops and wine. Farther down the road, we visited a mineral spring that had once been the site of a popular local health spa. It had closed some years earlier, and all that remained was a derelict building, with a bit of stonework around the pool. As we'd all brought bathing suits, we stopped long enough to enjoy a brief but rejuvenating dip.

Once we were on the road again, we questioned our guide as to why the spa had closed. Choosing his words carefully, he explained, "There was an uprising in the early 1980's, when Kanaks drove most of the white people off the east coast of the island. At the time, there were numerous deaths and two policemen were beheaded." He then continued, "French authorities reacted quickly, sending in a squad of commandos, and there were more deaths. An accord was eventually signed, agreeing that a referendum on independence would be held in ten years' time."

We were shocked to learn of the island's recent violent past, which was so at odds with the views of peaceful countryside that we were enjoying. When we then stopped at an east coast beach and toured a Kanak village, we wondered whether anyone lived there, as we saw no sign of villagers. Later, when we drove through another community, our guide asked us to close the windows, explaining that people occasionally threw rocks.

With our curiosity aroused, we spent the rest of our time in Noumea learning about the country's recent history. We first learned that the current population was 44% Kanak and 35% European, with most of the rest being immigrants from other French colonies. We then discovered that an independence movement had existed for many years, and in the early 1980's, the struggle had become increasingly violent.

Events came to a head in 1984, when a group of white settlers reportedly ambushed and killed ten Kanaks. That event brought further violence and more murders, until the country was on the brink of civil war. In January of 1985, the authorities declared a State of Emergency, but the pot continued to simmer for three more years. In April of 1988, members of the independence movement took thirty-five hostages, all but one being members of the gendarmes and the military.

The French sent in commandos, who successfully freed the hostages. But in the process, nineteen Kanaks were killed, as were two members of the hostage-recovery team. Amid charges that many Kanaks died while in military custody, a protocol for an agreement was negotiated and signed in August 1988. In it, the government agreed to transfer more powers to three regional assemblies and to hold a referendum on independence in 1998.

In fact, that referendum was held and approved the Noumea Accord. The accord gave more autonomy to local government and provided for another referendum to be held before the end of 2018. The population will then vote on whether or not to become fully independent.

A few days after touring the island, we treated ourselves to lunch in a downtown restaurant overlooking a park. The afternoon was pleasant and the food and drink excellent, so we were quite enjoying ourselves. But I again noticed something that I'd observed earlier in the week. The benches in the park were filled with white faces enjoying the sunshine, while dark faces sat on the grass in the shade of the trees. Although we had previously noticed that the two cultures didn't mix, we now recognized the tension, with a highly visible police presence in the downtown area on Friday and Saturday, and actually, whenever people gathered in any numbers.

By then, we were well into our final preparations for the passage to Australia. Dave had only to service the engine, while I was putting the final stitches in the courtesy flag. We had studied charts and pilots, paying particular attention to the approach to Brisbane, as rumor was rife about the sand bars littering Moreton Bay. Also, the buoyage system in Australia was the opposite of what we'd learned, so we kept reminding each other that the rule of "red right returning" had become "red right reaving".

We didn't think weather would be a problem on route, although a chap had been walking up and down the dock with a weather fax when we first arrived. Showing it to anyone who would look, he had worriedly pointed out that a cyclone was developing to the northeast. We weren't unduly concerned at the time, but I did wonder if the El Niño weather pattern might bring an early start to the cyclone season.

We purchased only a few provisions for the crossing, as we wanted to use up our existing supplies; Australian Agriculture was very strict about what they allowed into the country. We then prepared the boat for sea and made the trek down to the building that housed Customs, Immigration, and Harbor Master to obtain our departure clearance.

CHAPTER 27

Passage to Australia

Early on Sunday, October 19, Dave tuned in the cruisers' net on HF radio and heard reports of light winds at sea. With the weather forecast also calling for light winds, we hesitated, wondering whether we should leave. However, we could see no point in sitting around the marina so ran up to the office and paid our bill, then motored across to the fuel dock for diesel. By the time we set off for the north channel of the Passes de Boulari, it was almost noon.

Relaxing in warm sunshine, we motored through sheltered waters for the next three hours. It didn't seem possible that we were now starting on the last leg of our voyage. We expected to be at sea for nine days, as our route from Noumea to Brisbane would take us west-southwest for 840 nm. Boisterous SE trade winds had made the route difficult earlier in the season, but they were now subsiding. In fact, during the coming months, north winds would occasionally blow down the Australian east coast, bringing hot, humid weather and sometimes a cyclone.

As we approached the pass through the reef, a sudden squall brought strong winds and rain, reducing visibility to zero. Dave just steered *Windy Lady* out into open water, and we stopped to raise the mainsail once the storm had passed. With SSW winds at 10 kt and a low two-foot swell rippling across the water, we then headed out to sea. In front of us, a huge expanse of open sea and sky beckoned us onward.

The winds eased during my sunset watch, then gusted up to 20 kt a couple of times and died. With the air still, the water flattened out, and in the ensuing silence, the sound of a jet engine had me searching the sky overhead for an airplane. The heavens then filled with stars and *Windy Lady* drifted on into the night. An hour later, a quarter moon climbed up out of the sea, shining its light across the water, and soon a weak 5-kt breeze rose out of the south. With a boat speed of just over two kt, our progress was painfully slow, but our berths were then as comfortable as any bed ashore.

The winds increased to 5–10 kt shortly after midnight, and during the next twelve hours, we sailed fifty nm, averaging just over four kt. The temperature dropped overnight, however, and Dave had pulled on both sweater and rain jacket before daybreak. By noon, winds were down to 5–7 kt and swells were up to three feet. In our first twenty-four hours at sea (Day 1), we logged 90 nm.

While winds remained light, seas grew higher, and within an hour, long, regular southerly swells, six feet high, were sweeping beneath the keel. They knocked the wind from the sails, so *Windy Lady* rolled and twisted, and Dave spent the afternoon fighting to keep the bow on course. An hour into my sunset watch, winds picked up to 7–8 kt; although not much of an increase, it made all the difference. The sails now held the wind, so I no longer had to fight the helm, and the boat speed went from three to four kt.

The breeze died away early in Dave's 2000-watch, and he again fought to keep the boat moving forward. Three hours later, winds rose out of the NNE at 10 kt, having shifted over 150 degrees. As they were now on the stern, he tacked and reset the sails, but the boat wallowed in four-foot swells, so he still struggled with the helm. An hour into my midnight watch, winds backed to the north and 15-kt gusts started turning the bow into wind, making it even harder to keep on course.

When Dave relieved me at 0400, I suggested putting a reef in the mainsail, but he decided against it. A line of squalls then struck out of the darkness, bringing 30-kt gusts and driving rain; in between, winds were light and the boat rolled and twisted relentlessly. One squall backwinded the sails and with rain teeming down, he waited for it to pass before bringing the boat back on course. His comment in the journey log reads, "A night from hell!"

At daybreak, the bobbing lights of a freighter passed by one nm off the bow; soon after, the breeze began to back and we then left the squalls behind. By 0800, winds were steady at 10–12 kt from the west and seas were down to three feet. But at noon, seas were eight feet high and winds were gusting at 12–17 kt from the south. Our progress was depressing and we recorded only 85 nm for Day 2.

An hour later, winds were gusting from 10–20 kt, and the day had turned cloudy and cold. Although we were close to the Tropic of Capricorn, it felt like we'd left the tropics far behind. Seas were down to three feet when I came on watch at 1600, but conditions seemed unsettled, so we put a second reef in the main and then headed into a cool, misty night.

Just after 2000, winds settled at 15 kt from the SSE, and for the first time during the passage, Dave made twenty nm during a watch. But winds gusted from 7–28 kt during my midnight watch, and I fought with the helm and made only thirteen. Winds then veered to the south, gusting from 20–25 kt, and we sailed forty-six nm over the next two watches. At noon, we recorded 114 nm for Day 3.

That afternoon, Dave rode the wind and ten-foot swells and logged twenty-seven nm. I was a little more cautious going into the night and reefed in the headsail, so only made twenty-four. The winds then eased slightly, as did the seas, and we continued to make good time overnight. Dave pulled the reef out of the headsail at 0400, but ran into squally conditions and had to babysit Otto. About midmorning, the squalls disappeared, as winds backed to the SE, settling at 12-15 kt with seven-foot seas. At noon, we recorded 136 nm for Day 4.

The winds moved constantly over the next twelve

hours, veering from southeast to south and back again, while gusting from 15–25 kt. Seas were rough, so we weren't getting any sleep, and Otto needed constant attention. Winds dropped to 10–20 kt during my midnight watch and when seas became noticeably quieter, I started thinking longingly of my berth. When Dave relieved me at 0400, I think I was asleep before my head hit the pillow.

Two hours later, I was jerked out of a deep sleep as *Windy Lady* again started heaving and twisting. Trying to cushion myself against the boat movements, I pretended to ignore the sounds of flogging sails and hammering in the lockers. But after forty-five minutes, I flew into a rage. Storming out to the cockpit, I screamed, *"Furl the g-- d--- sails! Do something! Anything!"*

There was nothing to be done because the wind had died, leaving *Windy Lady* at the mercy of four-foot swells. Still, Dave did what I asked and dropped the sails. The boat immediately turned broadside to the swells and rolled over so far that I was practically thrown from my berth. Admitting defeat, I wearily got up and started tracking down the knocking sounds coming from the lockers, securing those items that had worked loose.

With winds light and variable, I didn't even try to steer a heading during my morning watch. I just turned the bow as needed to keep wind in the sails. When the wind died, leaving *Windy Lady* rolling in the swells with her headsail flogging, I furled in the sail and waited. Eventually, the wind picked up from the SE, so I pulled the sail out again and slowly the boat moved forward.

Dave got up soon after, and at 1045, started the engine. For Day 5, we logged 117 nm.

We ran the engine for six and one-half hours, leaving the steering to the autopilot. As soon as the water was hot, we luxuriated in badly needed baths, after which I mopped up the cockpit. As we relaxed, soaking up the sunshine, I realized how much of a trial the passage had become.

At that moment, a troupe of dolphins appeared, four of them tail walking towards us. In the clear water, I could see the shadowy outlines of others, some 100 feet away. The pod stayed maybe ten minutes, racing back and forth alongside the boat, turning 180 degrees in the blink of an eye. As always, they cheered us up, and I was still smiling long after they disappeared.

Just before 1600, a freighter passed by half a mile to port, and Dave called it on VHF radio. A crewmember from the Meridian Slobovia responded but didn't speak any English, so the conversation was brief. The winds then stirred out of the SSE, and at 1730, with winds at 7–12 kt and three-foot seas, we shut down the engine and started sailing again.

As usual, I studied the sea as the light faded from the sky and noticed again the two pairs of small black and white birds that had accompanied us for the past few days. I had decided they must be a type of storm petrol. The birds flew low over the sea, skimming the surface of the waves, and their white markings stood out clearly against the dark waters. They appeared to dip one foot into a wave before soaring up and away. I was sure they were feeding, but had no idea on what. The birds disappeared as the darkness deepened, and I then counted the stars as they appeared.

A storm front moved through before midnight, bringing ESE winds of 8–15 kt. A number of squalls followed, with 30-kt gusts and rain; in between, the winds dropped below 5 kt. As Otto couldn't cope, we hand-steered throughout the night, but we both managed to get a few hours' sleep.

By first light, winds were steady at 7–10 kt but they moved constantly between ESE and SSE. Dave then set the autopilot, which worked well in these conditions. About 0900, the wind died, leaving us sitting 150 nm off the coast of Queensland, Australia. With patience running thin, he started the engine. We logged 111 nm for Day 7.

After motoring through three-foot seas for almost twelve hours, we set the sails again at 2030 with SE winds at 7–10 kt. Next morning about 0630, the winds backed to the east and dropped to 7 kt. Dave then set the sails wing-on-wing, hoping to keep the boat moving in the right direction. At 1100, tired and frustrated by the continuing struggle, he started the engine. For Day 8, we logged 103 nm. With light winds and seas between three and five feet, we motored for the last twenty-four hours of the passage.

We had hoped to be off Calundra Head at first light, intending to pick up the shipping channel and follow it through Moreton Bay's shifting sandbars. Instead, we arrived in the middle of the night, so both kept watch from 1600 onward. Dave sent me below at midnight, saying, "We're going to be a while yet, so you go down and get some rest. I'll call you when we get closer to shore." He then added, "But first, find a waypoint that I can enter into the GPS to guide us in."

Looking in the Queensland Coast pilot, I now made an error that could have been disastrous; I gave him a position that was actually on land. For his part, Dave blindly followed it in until he was within spitting distance of shore and in only seventeen feet of water. When he called me up to the cockpit at 0130, I couldn't believe my eyes. Lights blazed across the shoreline in front of us, while waves broke on rocks to seaward.

Suddenly wide-awake, with an icy pit growing in my stomach, I checked the chart and saw that we were inside Bray Rocks. I then stood at the rail, my eyes probing the dark waters for signs of danger, while Dave monitored the depth gauge and steered *Windy Lady* back to safety. After making our way to a GPS waypoint near the entrance to the shipping channel, we finally identified the red and green lights that marked it. Until then, the lights on shore had concealed them.

The shipping channel wasn't without danger, as the bow wave from a freighter had reportedly damaged a yacht, but it still seemed the safest route across Moreton Bay. I was at the helm as we started in, and made sure to stay well over on the starboard edge. Before long, we noticed a couple of small lights moving toward us, but they didn't look right for a ship. The dimly lit vessel was moving quickly though, and we were making six kt on a flood tide. Suddenly, the dark bulk of the freighter loomed over top of us and then it was gone, leaving *Windy Lady* rocking in the bow wave.

Feeling intimidated, I checked frequently in both directions for traffic but saw only the vessel moving away from us. When I looked back five minutes later, another huge, dark mass was steaming up behind. I barely had time to tighten my grip on the wheel, before *Windy Lady* was again rocking in a bow wave. The ships moved fast and were unsettling, but we never strayed from our self-imposed position on the edge of the channel, so never had a problem.

The channel markers were up to five miles apart but were easy to spot, and we had no problem staying on course. When daylight came at 0530, we were three-quarters of the way through, and several more ships had passed us by. We now left the shipping lane and navigated across the width of Moreton Bay. More concerned about sandbars than ships, we stayed well out from shore as we entered Deception Bay.

Our chart showed only one set of marker buoys leading into shore, so we turned into the first set we saw. A minute later, a Customs Officer was calling on VHF radio, warning that we were heading into shallow water and should look for a second set of leads. Reversing course, we then found the correct ones, and Dave steered *Windy Lady* into Scarborough Marina. At 1030 on October 27, he brought her alongside the Customs dock, and I threw our mooring lines across to waiting hands.

When the Customs and Immigration Officer climbed

aboard, he immediately demanded a cup of coffee. He then thumbed through a pile of forms, explaining that he was expecting sixteen boats to clear in that day (eleven showed up). His paperwork was quickly completed, and we then sat and waited for the Agriculture Inspector. He showed up several hours later, but within five minutes of his arrival, I was staring in dumbfounded disbelief, having no idea what had hit us or why.

He went through every container in every locker in the galley, opening each one and looking inside. He shook the contents about, checking for bugs, but didn't bother to put the lids back on properly, so spilled rice all over the place, including the tracks for the sliding doors. He didn't say a word, just piled containers high on the galley table as he emptied the lockers. Leaving the galley in a shambles, he took some dried beans, butter, and eggs when he left; all of these had been in plain sight when he arrived.

After moving over to a berth on a nearby finger, Dave put the boat to bed while I restored order to the galley. It was late afternoon before we were free to relax in the cockpit and discuss the passage. We logged 863 nm during our nine days at sea, not much more than our rhumb line course of 840 nm. But with six-foot swells and no wind, the crossing had often been frustrating and difficult.

We spent the next few days deciding where to berth *Windy Lady* and began talking about exploring Australia. Meanwhile, we spent hours walking through the communities of Scarborough, Redcliffe, and Kippa Ring. We had landed in a very beautiful area on Moreton Bay, and much appreciated the friendliness and helpfulness of the locals. Within days, however, hot, humid tropical weather rolled down the east coast. We then made our first lifestyle adjustment, giving up coffee, as just one sip caused the sweat to pour off our bodies.

Within a few weeks, Dave bought a 1987 Ford Falcon station wagon, and we set off to explore Australia. Along the way, we stopped in Mount Isa, where Dave had worked in a mine when he immigrated in 1962. We also stopped in the small town of Coonamble, where I worked on a sheep station in 1968. These experiences had undoubtedly provided the impetus behind our desire to sail to Australia.

I don't think we appreciated the fact that we had achieved our dream because the dream had become our reality. In a period of 507 days, we spent 102 days at sea, sailing 10,300 nautical miles (19,075 km). That experience changed us forever. The truth was that we both enjoyed living on the boat, enjoyed the adventure, but most of all had tasted a freedom that we never knew existed.

Although we hadn't thought about what would come next, there was a whole world out there to explore. So much more to see and do, and we saw no reason to quit; in fact, we would live on *Windy Lady* for another thirteen years. We continued to sail for five of those years, but when the world changed with "9-11" in 2001, so did our plans; we then traveled by plane, by car, and on foot.

We visited the Delphi in Greece, Machu Picchu in Peru, and Angkor Wat in Cambodia. We hiked up to 18,000 feet in the Himalayas and traveled the Silk Route through China. We peered down into a section of the

Cuchi Tunnels in Viet Nam and stared with disbelief at the site of mass graves in the killing fields outside Phnom Penh in Cambodia. There were exceptional evenings spent with people we'd just met and would never see again, while we enjoyed semi-permanent relationships with others.

Many more adventures awaited us—and there is still so much to tell....

Glossary

aft, abaft: near or toward the stern

autopilot: self-steering mechanism used under power

backwind: to deflect air onto the back of a sail

beam: the width, or widest part of the deck

beam reach: fastest point of sail, with wind coming over the beam

beamy: broad beamed

beat: point of sail, with bow held as close to the wind as possible

bilge: lowest point inside the hull

binnacle: the stand in which a ship's compass is kept

blanket: to prevent the wind from filling a sail

boom: the horizontal spar that supports the foot the mainsail

bow: forward part of a boat

broach: sudden, sharp turn caused by wind or sea; can cause a boat to capsize *broad reach*: point of sail, with wind coming over the quarter

broadside: beam on to either wind or sea

bulkhead: a support below deck that strengthens a boat

buoy: a floating aid to navigation

chain plate: a metal plate used to fasten a stay to the hull

chandlery: a store selling boat supplies

coaming: the raised edge around the cockpit that keeps water out

cockpit: an opening in the deck from which a boat is steered

companionway: stairs between cockpit and cabin

course: direction in which a boat is steered

dodger: a spray shield that protects the cockpit

easting: distance traveled towards the east

fender: a cushioning object placed between boat and dock

flog: for a sail to flap or flutter when no longer supported by the wind

foredeck: deck area between mast and bow

forestay: supporting cable running from upper mast to bow

freeboard: distance from top of side to waterline

galley: kitchen area of a boat

gimbal: swing support that allows a stove to stay level in rough seas *grab rail*: hand-hold fitting mounted inside the cabin

HAM: restricted frequencies on an HF radio

HF radio: high-frequency radio used for long distances

hank: fastener affixed to the staysail that attaches to the inner forestay

hatch: an opening in the deck that can be sealed off

head: a marine toilet, or the room in which it is located

heading: the direction in which the bow points at any given time

headsail: sail attached to the forestay

heel: the angle of the boat to the water

helm: the wheel controlling the rudder

hove-to: to have sails/helm positioned so that a boat remains almost stationary

hull: actual body of the boat

jack line: a safety line running down the length of the deck

jibe: to turn a boat so that the stern passes through the eye of the wind

keel: an extension of the hull that goes deeper into the water

km: kilometer - equals .54 nautical mile

kt: knot - equals one nautical mile/hour, or 1.852 km/hour

lee, leeward: the side sheltered from the wind

lifeline: a cable running along the edge of the deck supported by stanchions

mainsail: principal sail on the main mast

mast: a pole on a boat that supports the sails

mb: millibar (a unit of atmospheric pressure)

moorage: a place to secure a boat to shore or sea bottom

nm: nautical mile - equals one minute of latitude, or 1.852 kilometers

pad eye: metal eye permanently secured to the boom

painter: a line attached to the bow of a dinghy

pooped: to have a wave break over the stern, can cause a boat to capsize

port: the left side of a boat when facing the bow

pulpit: guardrail around the bow

push pit: guardrail around the stern

quarter: the part of a vessel's side near the stern

reef, to: reduce the area of the mainsail using pre-established reefing points, or to partly furl the headsail

rhumb line: shortest distance between two points at sea

rudder: an underwater vertical surface that steers the boat

scupper: a drain hole in deck or cockpit

sloop: a sailboat with one mast and one headsail

sole: cabin or cockpit floor

spreader: horizontal support almost halfway up the mast

stanchion: metal post supporting the lifelines

starboard: the right side of a boat when facing the bow

stays: strong cables supporting the mast that run fore and aft

staysail: a small jib sail attached to an inner forestay

steerage: the effect of the helm on a boat, or the act of steering it

stern: back end of a boat

tack: to turn a boat so that the bow passes through the eye of the wind

thru-hull: a fitting providing a secure hole through the hull below the
waterline

toe-rail, rail: the outer edge of the deck, usually raised

transducer: electronic device needed to measure water depth, boat speed, etc.

transom: the aft wall of the stern

turnbuckle: devise for adjusting tension on the stays

VHF radio: very high frequency radio, used for local communication

westing: distance traveled towards the west

whisker pole: a pole used to hold a sail out in light winds

windage: wind resistance of a boat

CPSIA information can be obtained
at www.ICGtesting.com
Printed in the USA
LVHW05s0026120918
589883LV00001B/1/P